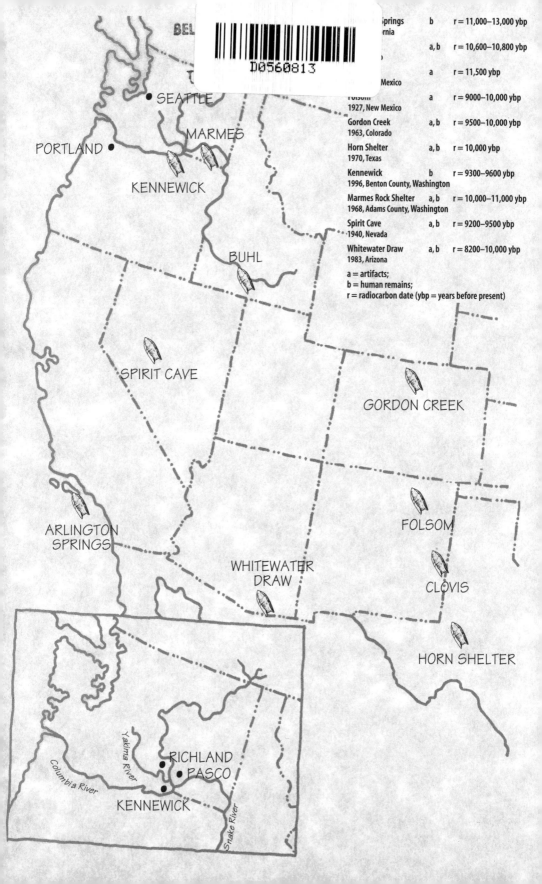

BEL...

D0560813

SEATTLE

MARMES

PORTLAND

KENNEWICK

BUHL

SPIRIT CAVE

GORDON CREEK

ARLINGTON
SPRINGS

FOLSOM

WHITEWATER
DRAW

CLOVIS

HORN SHELTER

RICHLAND
PASCO

KENNEWICK

Columbia River

Yakima River

Snake River

...Springs   b   r = 11,000–13,000 ybp
...ornia

...o   a, b   r = 10,600–10,800 ybp

...Mexico   a   r = 11,500 ybp

Folsom   a   r = 9000–10,000 ybp
1927, New Mexico

Gordon Creek   a, b   r = 9500–10,000 ybp
1963, Colorado

Horn Shelter   a, b   r = 10,000 ybp
1970, Texas

Kennewick   b   r = 9300–9600 ybp
1996, Benton County, Washington

Marmes Rock Shelter   a, b   r = 10,000–11,000 ybp
1968, Adams County, Washington

Spirit Cave   a, b   r = 9200–9500 ybp
1940, Nevada

Whitewater Draw   a, b   r = 8200–10,000 ybp
1983, Arizona

a = artifacts;
b = human remains;
r = radiocarbon date (ybp = years before present)

Riddle of the Bones

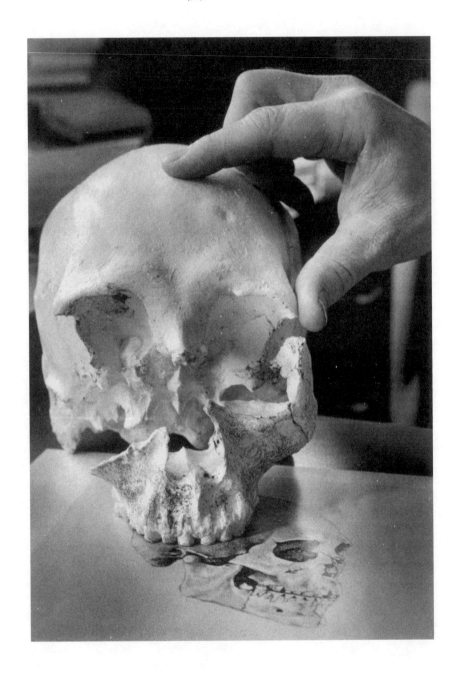

RIDDLE

OF THE

BONES

Politics, Science, Race, and
the Story of Kennewick Man

ROGER DOWNEY

COPERNICUS
AN IMPRINT OF SPRINGER-VERLAG

Frontispiece: Cast of the Kennewick Man skull
made by Jim Chatters. Below the cast is a
drawing of same made by Claire Chatters.
(Photo by Jerry Gay.)

979.7
D

© 2000 Roger Downey

Published in the United States by Copernicus,
an imprint of Springer-Verlag New York, Inc.

Copernicus
Springer-Verlag New York, Inc.
175 Fifth Avenue
New York, NY 10010

Library of Congress Cataloging-in-Publication Data
Downey, Roger.
    Riddle of the bones : politics, science, race, and the story of Kennewick Man /
Roger Downey.
        p.   cm.
    Includes bibliographical references and index.
    ISBN 0-387-98877-7 (hc. : alk. paper)
    1. Kennewick Man.   2. Human remains (Archaeology)—Washington (State).
3. Cultural property—Repatriation—United States.   4. Indians—Origin.
5. Washington (State)—Antiquities.   I. Title.
E78.W3D69   1999
979.7'01—dc21   99-040567

Manufactured in the United States of America.
Printed on acid-free paper.

Text designed by Joan Greenfield.
Cover map illustration by Dugald Stermer.

9 8 7 6 5 4 3 2 1

ISBN 0-387-98877-7   SPIN 10730314

*This book is for Peter Ward, who thought
it should be written, and for Knute Berger,
who made it possible to write it.*

CONTENTS

PREFACE

The more homogenized human culture becomes, it seems, the deeper interest individuals take in what makes them unique: nationality, creed, family heritage. When archeologists turn up fresh evidence of the whereabouts and doings of our remote ancestors, it is perfectly natural that the story turns up on the evening news.

But even the most striking discoveries rarely stay in the headlines long. Ancient bones and implements may be newsworthy, and the older they are, the more so. But sheer age, though interesting, may take the edge off the excitement. The older they are, the more common, too, the heritage of the species they represent, and the less they reinforce our individual sense of identity. When a blue-ribbon scientific panel returned from southern Chile in the spring of 1997 to vouch for Tom Dillehay's controversial discovery of a rich, complex, 12,000-year-old occupation site there, the media buzz barely lasted a week.

The discovery known as "Kennewick Man"—the fairly complete skeleton of a fifty-ish male that turned up in the summer of 1996 in a small town in eastern Washington state—is a remarkable exception to the rule. More than three years after his remains washed out of a bank on the Columbia River, K-Man keeps popping up in the media, the subject not only of thousands of newspaper articles published round the world, but of multiple lengthy television documentaries as well, while the scientist who first publicized the skeleton's importance has come to be perceived as a kind of Indiana Jones figure, receiving the ultimate ratification of celebrity status, a profile in *People* magazine.

Why? Not because of the intrinsic scientific importance of the find: Even if the skeleton ultimately proves everything its finders claim for it, its character and the manner of its discovery ensure it no more than a minor place in the history of human occupation of the Americas. It is not their intrinsic value as evidence that makes these bones newsworthy; rather the spin applied in disclosing that evidence.

In the media, the saga of Kennewick Man and Dr. James C. Chatters has played out as a rerun of the battles between Charles Darwin and Bishop Wilberforce following the publication of *The Origin of Species*, between Clarence Darrow and William Jennings Bryan for the hearts and minds of the jurors assembled to settle the fate of John Thomas Scopes, and as a straightforward confrontation between science and superstition, the individual's right to pursue knowledge freely blocked by bureaucracy in service of a reactionary religious agenda.

As usual, the reality behind the familiar scenario is far more complex and ambiguous. Kennewick Man's story embodies not only a conflict between science and religion, but conflicts between scientists espousing radically different worldviews. On close reading, it exposes a jagged faultline, largely generational, between investigators still wedded to a traditional approach to investigating the human past, and others who embrace sophisticated techniques of investigation unknown to their teachers' generation.

A closer view also makes it plain that superstition and rigid ideology are by no means restricted to one side in the Kennewick controversy, and that personal and institutional pressures and ambitions have at least as much to do with the stance adopted by some of the scientists involved as any abstract devotion to truth. Even hard scientists today admit that there's a certain amount of truth in the "post-modern" notion that absolute truth is at best an ideal, not an attainable goal, that even the elegant mathematical formulations of quantum mechanics can't claim any ultimate validity, but partake to some extent of the nature of myth—of the stories we tell ourselves and each other to explain who we are and what role we play in the grand scheme of things.

But myths come in more and less sophisticated versions, and narratives that purport to reflect reality differ in their tolerance for complexity and contradiction. In this book, I have tried to use the often sad, often comic, frequently farcical saga of K-Man to explore several parallel narratives: how a fifty-year-old picture of early human life in

North America that has shaped the dreams of generations of scientists and schoolchildren is slowly eroding under the assault of new information; how the changing economic and political status of academic science is transforming the fields of anthropology and archeology in the United States to the intense discomfort of those once favored by the existing system; how technological progress has interacted with societal change to produce a new breed of investigator and a new way of looking at human prehistory; and how all these changes have been furthered, impeded, warped, and exploited by the media, largely to the detriment of both scientific progress and public understanding.

*A*s a nonspecialist, I have had to depend heavily on others for guidance in and evaluation of research in a dozen scientific fields. I have tried, in the notes to each chapter, to give credit for their assistance, while making as clear as possible where their opinions and interpretations of the data cease and mine begin. In my attempts to escape journalistic cliché and to understand how scientific research actually gets done, I owe a great deal to three scientists who are also superb historians: personally, to Donald K. Grayson of the University of Washington and David J. Meltzer of Southern Methodist University, both of whom assisted me directly in my research; and to the example set by Stephen Jay Gould of Harvard University, who for 25 years has hammered home, in essay after brilliant essay, the essential, oft-forgotten fact that science is an assemblage of artifacts—of things made by human brains and hands—as contingent on history, individual character, and chance as any other form of creative activity.

River

A Day at the Races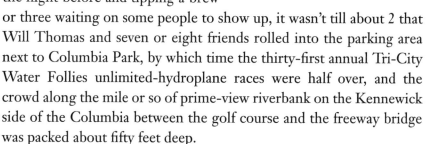

The original idea was, get together Sunday morning at
Ryan Hickey's place in Richland,
then everybody hop in Bill Ashley's
pickup, and head on down to the
river and beat the crowd. And, with
any luck, before they got the ticket
booth open. But what with the party
the night before and tipping a brew
or three waiting on some people to show up, it wasn't till about 2 that
Will Thomas and seven or eight friends rolled into the parking area
next to Columbia Park, by which time the thirty-first annual Tri-City
Water Follies unlimited-hydroplane races were half over, and the
crowd along the mile or so of prime-view riverbank on the Kennewick
side of the Columbia between the golf course and the freeway bridge
was packed about fifty feet deep.

Most of Thomas's pals resignedly got in line to pay their $15 a head
to see what was left of the races, but Thomas, a Water Follies veteran,
wasn't about to give in so easily. If the front door was guarded, there still
might be a way around the side. Persuading a couple of his buddies to
show a little initiative too, and stashing a cold brewskie or so under the
waistband of his cut-off sweatpants, he started picking a way through the
shrubbery choking the riverbank between parking lot and golf course.

It was pretty nasty going. The temperature was close to 100 degrees, high even for the Tri-Cities at the end of July, and on a windless

afternoon next to the river the humidity was verging on 100%, too. The ground was mushy. One of the seeps draining the bluff to the south merges with the river there; marsh grass and cattail choke the few patches of ground not blocked by the whippy grey-green tangle of Russian olive.

After five minutes of pushing through the undergrowth along imaginary paths, the three found themselves back beside the river road, closer to the ticket gate than they started. "Screw it," Joe Wick said to his friends, "I'm going to pay," and he trudged back to join the group in line, brushing the twigs out of his hair.

With only Dave Deacy as companion now, Thomas struck back into the bushes. This time they worked their way through to the river, found a spot where they could clamber down the crumbly clay of the eight-foot bank to the edge of the slow-moving water. Once by the riverside, success was in sight. Though it's frowned on by the authorities, there's no actual law against taking to the water and wading downstream for a closer view of the race. Anyway, with the final heat yet to be run, and with 20,000 law-abiding citizens packed like punks in a mosh-pit on the bank above, the authorities had more important things to occupy them than a couple of wading teenage gate crashers.

The river itself was no problem. At this point, the Columbia is a good half-mile wide, but so shallow that you can wade out a hundred feet or more without the water getting much higher than your knees. Streamflows the previous winter had been the highest in a hundred years—so high that the Army Corps of Engineers had to open the spillways at McNary Dam forty miles downstream to keep the rising waters from flooding the million-dollar homes that line the bank in Pasco, north of the river. But now the flow was sluggish, just fast enough to keep a fine mist of silt in suspension above the muddy bottom.

It was the gluey, adhesive mud along the bank that sent Thomas wading out into the river, looking for more solid footing, while Deacy stuck close to the bank, though it meant clambering over roots and ducking under overhanging branches. Even twenty feet offshore the bottom wasn't much better, but the water was at least clear enough to see where you were putting your feet, and when you're barefoot and picking your way over a bottom littered with the detritus of a throw-away civilization, that counts for something.

Thus it happened that, squinting through the two-foot-deep water for safe footing, Will Thomas saw what he thought was a smooth, round rock looming in his path, decided to screw with his pal Dave a little, and thereby secured a modest place in the history of science. "Hey," he shouted to Deacy, reaching into the water and plucking out the rock, "we have a human head."

And what do you know, that's exactly what they had: a complete human skull, river mud oozing from its eyesockets and a double arc of brown teeth grinning in the sun. For a timeless moment, the two young men stared at their find.

Abruptly, the delicacy of their situation came home to them: in the river, toting beer, one of them under age, and in possession of a human skull.

"Just put it back where it was," was Deacy's first thought. "No, we gotta do something," Thomas responded. But what? Struggle back through the brush, find a cop, and report the discovery? In the middle of the last day of Water Follies? With Dave Villwock's "PICO American Dream" already warming up to take on Mark Tate's "Smokin' Joe's" for the Budweiser Columbia Cup? "No way," said Thomas.

Well, what then? About sixty yards downstream some younger teens had escaped the crowd to play in the water. Had they noticed anything? "Screw 'em," thought Thomas, "I found this thing." Skull tucked under his arm, he waded with Deacy downstream along the bank until they found a shadowy spot under the shrubbery where tall weeds were growing. There they stashed the skull among the weeds and returned to the real business of the day: seeing Villwock extend his summer-long winning streak, hitting an average speed of 149.447 mph and kicking Tate's butt by a good couple of roostertails.

By the time Thomas and Deacy returned to reclaim their grisly prize and clamber back through the underbrush, the parking lot had cleared enough to leave their law-abiding friends room to toss a frisbee around to pass time until their return. "Hey look, I found a skull," Thomas shouted. "You're full of crap" quickly changed to "Whoa, where'd you get that?" The owner of the truck, a carpenter, emptied a five-gallon plastic paint tub of miscellaneous tools and offered it to Thomas as a carrying container, and the entire band set out looking for a cop.

They didn't have to go far. Water Follies is the biggest event of the Tri-Cities year, drawing visitors from all over eastern Washington state, northeast Oregon, and northern Idaho, and every professional and volunteer police, fire, and emergency team deploys a full complement to handle the crowd. Compared to past years, the 1996 Follies had been almost disappointingly tame. A vigorous campaign to stamp out drunkenness and rowdiness had been so successful that, apart from the usual cases of dehydration and heat stroke, the afternoon's major trouble calls had been to rescue two dachshunds from hyperthermia in a locked car and pump up a baby stroller's flat tire.

For Cal Nash, a police officer on loan for the day from nearby Richland, a skull in a white plastic bucket made an agreeable diversion from standing in the broiling sun directing people to the nearest Porta-Pot. Thomas and his friends were soon outnumbered by policemen from all three Tri-Cities and the Benton County sheriff, every one of them barking importantly into walkie-talkies and cell phones.

One of the calls brought a county patrol craft, on hand in case, as often happens, a hydro flips its driver into the water. By now a Kennewick officer, Keith Sharp, had taken custody of the bucket from Nash, only to be relieved of the burden in his turn by Detective Sergeant Craig Littrell. Lines of authority established, Littrell ordered Thomas into the patrol boat to guide the officers to the discovery site. Thomas, already living dangerously by openly sipping a brew from the blue plastic cooler in the back of his friend's truck, said he wouldn't go unless co-discoverer Deacy was allowed to come too. After a tense moment, Littrell assented. Tucking another bottle into the waistband of his sweats, though covering it with his shirt as a token concession to propriety, Thomas swaggered aboard the boat with Deacy, the envy of his peers.

It took less than a minute to cover the distance Thomas and Deacy had so laboriously waded. Despite the low-angle sun (it was after 7 P.M., but still an hour and a half before dark), it took only another minute for informed cop eyes to spot bones everywhere, some unquestionably human, including one chunk of thighbone that Deacy had picked up and tossed away, thinking it was a stick. Littrell heated up his cell phone and called the county coroner to tell him his services were required in Columbia Park. Then he placed another call to the volunteer Columbia Basin Dive and Rescue team (there, like the patrol boat, to help luckless hydro drivers), summoning them to the scene to scout for further remains, meantime ordering his own team to tape off the area as a possible crime scene.

It was only on the way back to the landing that one of the officials took the opportunity to have a few quiet words with Thomas on the subject of cooperating with police officers in the execution of their duty, public consumption of alcoholic beverages, and the legal drinking age in the state of Washington. Thomas and Deacy were somewhat subdued by the time they arrived back at the parking lot, but cheered up when John Stang, a reporter for *Tri-City Herald* who heard the report of a skull in the Columbia on the newsroom police-radio scanner, interviewed them about their adventure. Having squeezed as much excitement from the discovery as it seemed likely to yield, the whole gang piled into Ashley's truck and headed back to Richland for Miles Wierman's kegger.

Sure enough, the next day in the *Herald*, there it was on page five, under the laconic headline "Skull found on shore of Columbia": very much a just-the-facts story, concluding with the words "So far, police have no idea of the age or origin of the skull. It will be sent to the state crime lab to be analyzed, Littrell said."

If the bones had been sent west of the mountains to the State Patrol crime lab in Lacey, odds are the world at large would never have heard of Kennewick, let alone Kennewick Man. But in Benton County, Washington, as in most jurisdictions, when police find human remains, fresh or well seasoned, a civilian expert has to pass on them. So among the calls Littrell placed on the brief boat ride back from the discovery site was one to the Benton County coroner, Floyd Johnson.

He caught the coroner far from the madding crowd, watching a ball game in air-conditioned comfort at his Kennewick home. But it wasn't much of a game, and anyway, Johnson had been looking for a chance to try out the flashers on his brand-new Jeep Cherokee, so despite the apparently routine nature of the call—the Tri-Cities average eight or nine drownings a year, as well as their share of suicides and murder victims—he told Littrell he'd be right down.

Looking at the discolored skull and jawbone, Johnson was inclined to agree with Littrell: If this was a crime victim, the statute of limitations had run out a long time ago. With traditional Native American burial grounds dotted here and there along a good hundred miles of Columbia shoreline, the bones were probably of even more humdrum provenance, to be routinely examined and routinely repatriated to the tribe historically associated with the site of their discovery—in this case, probably, the Confederated Tribes of the Umatilla Indian Reservation, headquartered some seventy-five miles from Kennewick on the far side of the Columbia in Pendleton, Oregon.

As an ex-deputy in the County Sheriff's office, Johnson had no more formal training in medical forensics than Littrell, so since he was already here—and besides, the game would be over by now—the coroner decided he might as well take the bones straight off to someone qualified to vet them: his anthropologist friend Jim Chatters, who had often volunteered in the past to examine evidence submitted to Johnson's department. The plan had the added advantage of getting the remains into good hands immediately, because Johnson himself was leaving shortly with his wife for a couple of weeks in the California back-country. Anyway, Jim would want to know about the discovery right away. Jim was interested in anything to do with bones.

After calling ahead to make sure Chatters was home, Johnson hopped back in his Cherokee, the bucket on the passenger seat beside him, happily switched on the flashers his friends had given him such grief about, and hit the freeway west to Richland and Jim Chatters's split-level home out by the airport.

Just that casually was the process set in motion that led, exactly one month later, to the *Herald*'s John Stang attending a hastily convened

10 A.M. press conference in Kennewick City Council chambers. Assembled behind the table in the low-ceilinged, windowless, brick-walled room, a trio of familiar figures was lined up in front of the television lights: Coroner Johnson, solid and slow-talking as always; dapper, mustachioed Kennewick mayor Jim Beaver, taking a rare morning off from the sales floor of his furniture store; and between the two a small, intense man in a flannel shirt and jeans, with longish, thinning, straw-colored hair, wire-rimmed glasses, and a single small silver ring in his left ear: Jim Chatters.

Stang, whose beat included local Native American affairs, had run into Chatters before, when he was heading up the Hanford Nuclear Reservation's archeology and historic preservation program. Stang hadn't seen much of him since Chatters had left Hanford after some kind of dispute with the management and set up his own consulting service. With both Johnson and Chatters present, Stang figured there was archeology in the air, but it never occurred to him the story might be connected to the bones found in Columbia Park until Chatters, eyes downcast like someone unfamiliar with the glare of TV lights and talking in that reedy, intense voice of his, announced that the remains discovered the month before were both those of an old settler *and* those of a crime victim.

The male individual, a lanky five foot nine or thereabouts, and forty-five to fifty years old at the time of his death, had survived at least one murderous attack during his life, maybe two. At some point the man's rib cage had been crushed by a heavy blow and had healed imperfectly. That injury might have been an accident, said Chatters, but not the other: Embedded in the right wing of the man's pelvis was a two-inch shard of stone, not a formless chip such as might have resulted from incautious handling of blasting powder, but a recognizable spear-point, fashioned, as well as could be distinguished by X rays and CAT scans, in the Cascade stoneworking tradition: a type of flintknapping that went out of style on the Columbia plateau around 5000 years ago.

Some of the TV reporters whispered to their camerapeople to keep shooting and dashed out to find a phone to alert their producers to hold space on the noon news. But there was more—much more. Only the day before, Chatters announced, the University of California radiocarbon laboratory at Davis had called to tell him that a miniscule bone from the specimen's little finger had been calibrated at 8410 +/−60 "radiocarbon years." This meant that, after adjustment for the way the amount of carbon-14 in the atmosphere varies over the centuries, it appeared that the individual the sample was taken from had died over 9000 solar years before the present, making his remains among the oldest ever discovered in North America.

"Chatters said the number of North American skeletons or partial skeletons as old as or older than the Kennewick discovery 'can be counted on the fingers of one hand,'" Stang noted. "There are only three North American female skeletons at least that old." In those days, Chatters said, "women tended to have shorter lives than men because of childbearing. Now it's the other way around; we're under higher stress." Knowing male chuckles all round. All kidding aside, though: "This," intoned Johnson in one of his few contributions to the proceedings, "is one of the most unique anthropological finds ever made in this area."

Age, sex, and injury were only part of the information that Chatters's expert eye could glean from the remains, which "tell you a detailed story of his life. If you look at muscle attachments, for example, how built up they are," he told his rapt audience, "you can tell what kinds of activities that person performed in life." In this case the grooves and ridges left where muscle once met bone showed that "he didn't carry heavy burdens." They also showed that the man's left arm was seriously underdeveloped compared to his right, even withered, very likely as a consequence of the injury that crushed his ribs. The enamel of his teeth was badly worn; on the other hand, the man still had all his teeth at the time he died—something a lot of fifty-year-old moderns can't claim—indicating a diet with a good balance between soft and chewy foods, probably featuring a good deal of meat. Death, when it came, might have been due to an infection from the wound in his hip.

By this time, the atmosphere was less that of formal news conference and more that of swapping yarns round the old campfire. "He's got so many stories to tell," Chatters said. "When you work with these individuals you develop an empathy, it's like you know another individual intimately. This guy is a heroic guy, unbelievably tenacious in life."

All of this was meat and drink to the assembled press, but Chatters still had his ace of trumps in hand. Though 9000-plus years old, the remains didn't look like those of contemporary Native Americans, far from it. In fact, "he looks like no one I've ever seen before." But if he had to make a call, everything about these remains (relative proportions of bone to bone, shape and layout of the teeth, the overall conformation and subtle details of the skull) was markedly "Causasoid." Of all known varieties of humans, the Kennewick remains most resembled those of a "pre-modern European."

Stang, experienced in the way such language can ignite fires that are hard to put out, chose to omit the terms "Caucasoid" and "European" from his next-day coverage of the story, leading instead with the sheer age and the romantically battered condition of the remains. ("The wandering hunter was one tough hombre," Stang's account began. "A spear point driven into his hip couldn't stop him. Nor could a once-crushed chest.") Television reportage was less circumspect. Typical was the report from KVEW's Tyffani Peters, standing mike in hand with the sluggish Columbia in the background: "The skeleton is also a unique find because the man has European features, not Native American. Chatters says this find could mean North America's first residents may have looked very different from what we believed in the past."

Today the Tri-Cities, tomorrow the world. Within the week, hundreds of millions round the world had been informed that the skeleton of a 9000-year-old European had been found on North American soil.

Bones in the Basement 2

*M*ore than other sciences, archeology progresses at the mercy
of chance. A wading teenager comes
across a skull; a farmer out plowing
turns up fragments of pottery; a cowboy riding fences spots something
white in the side of a flood-cut arroyo. For every such happenstance
discovery reported, how many more
pass unrecorded, lost through incomprehension, indifference, or greed?

Maybe because the very stuff of archeologists' livelihood is so dependent on the vagaries of chance, a lot of people who take up the profession are, if not driven, at least remarkably focused personalities.
Think of Schliemann, calmly convinced that Homer's Troy lay open
for the digging beneath an unprepossessing hillock in Turkey, conjur-
ing away doubt and contrary evidence through sheer concentration,
determination, resolution.

"Kennewick Man" came to light by chance. What followed upon
his discovery would surely have been different, and certainly far less
riven by controversy, had the first examination of his remains fallen to
someone other than James C. Chatters, Ph.D. The more one learns
about what Hollywood calls "the back-story," the more the tale of
Kennewick Man comes to resemble a kind of epic tragicomedy, with a
protagonist seemingly shaped by fate to play the lead role in just such
a drama.

*I*f anyone was ever born with a head start in his profession, it was Jim Chatters. A lot of scientists who end up digging for a living—geologists, paleontologists, and archeologists among them—get started early, collecting arrowheads, agates, and fossils as a hobby. Jimmie was introduced to serious field work soon after he could walk, helping his botany professor father poke through coal mine slag-heaps for the rocky nodules in which fossils are often found. By the time Jim was 10, his father had changed professions, shifting from the slow-moving field of botany into the fast-growing area of radiocarbon dating, which had been developed by Cal Berkeley's Willard Libby just a year or two before Jim was born.

The job change entailed a move for the Chatters family from the University of Oklahoma to the southeastern Idaho town of Idaho Falls. The area around the National Reactor Testing Station where Roy Chatters worked was great fossil country; at 11, little Jim was able to spend a summer vacation week with a team excavating a prehistoric bison kill site on Birch Creek, near his dad's labs. Roy Chatters's next career move put him and his son near the center of the archeological action: In 1961 he accepted a professorship in the Geology Department of Washington State University, where he was to remain the rest of his life.

Located in the southeastern town of Pullman, deep in the rich, rolling dry-wheat Palouse country, Washington State rivals the University of Idaho less than twenty miles away for the title of most remote and countrified of all America's land-grant universities. But at the time Roy joined its faculty, "Wazzu" was no mere cow college. After the end of World War II, the United States Army Corps of Engineers found itself unemployed and devoted a large measure of its excess capacity to erecting dam after dam on the Columbia and Snake Rivers in the name of rural electrification, irrigation, and progress, incidentally drowning thousands of square miles of the Columbia Plateau and surrounding areas of Washington, Oregon, and Idaho.

Under pressure from scientists of the Smithsonian Institution, Congress grudgingly provided funds to survey the hundred of miles of riverbank shortly to be inundated for archeological remains. A young WSU anthropology professor named Richard Daugherty took advantage of this federal largesse to put together what was, for the time, a remarkable multidisciplinary scientific team. Daugherty's Laboratory of Anthropology included on its staff geologists, archeologists, a palynologist (a specialist in identifying pollens, essential as clues to past climate changes), and, to provide dates on old organic matter discovered by the team, radiocarbon expert Roy Chatters.

In childhood Jim had had a roughish time. Over thirty years later, his prime recollection of those years was of "constant bullying as the smallest kid in class."[1] Chatters didn't take the ill treatment passively; in high school he turned out for wrestling and developed a reputation among his peers as one not to mess with. From early childhood, he exhibited many of the traits supposedly typical of a "middle child": independence, stubbornness, intolerance of the often arbitrary shibboleths of the high school peer group. And being a loner if anything enhanced his commitment to science.

The Chatters family arrived in Pullman, Washington, just in time to participate in a dig that was to prove the most significant ever conducted in the Pacific Northwest: the excavation, between 1962 and 1969, of Marmes Rock Shelter. The site was a long, shallow groove in the cliff overlooking the Palouse River (a small tributary of the Snake) that threads its way through the dry plains and buttes that flank Washington State's lush dry-land wheat country.

Situated on the property of rancher Roland Marmes (pronounced "MAR-muss"), the site had many of the characteristics that Daugherty knew early inhabitants of the region had favored for temporary encampments. It was near a supply of life-giving water in a virtual desert, faced east to catch the direct rays of the low winter-morning sun, and was protected from wind by a sharp bend in the canyon upstream. Three seasons of summer excavation at the site quickly confirmed its archeological value. There was evidence of occupation 8000 years deep: everything from what was then the oldest known burial in the Americas to one of the dozen "peace medals" distributed by Lewis and Clark on their epic journey to the Pacific in 1804–1805; from the bones of prey species like elk, bighorn sheep, deer, and antelope and the projectile points that killed them to snail shells pierced for stringing that could have come only from the Pacific Ocean hundreds of miles downstream.

Rich as it was, Marmes was only one of scores of sites to be surveyed before erosion, irrigation, and dam building took their toll, and in 1964 Daugherty moved on, leaving Ph.D. candidate geologist Roald Fryxell behind to puzzle out the complex geology of the shelter and the riverbank below. Toward the end of the digging season, rancher Marmes offered to cut a trench through the thick sediments with his bulldozer to expedite Fryxell's labors. Walking behind the bulldozer in a narrow dust-choked cut twelve feet below the surface, Fryxell spotted a thumbnail-sized gray chip of broken bone—one of some two dozen disturbed by the passage of the dozer blade.

Only after two and a half years of seasonal digging at the site and painstaking analysis back in the Pullman labs did Daugherty and his team feel sure enough of what they had to make a public announcement: the cranium of a youth between 9000 and 13,000 years old, by far the oldest reliably dated human remains ever discovered in the Americas. On April 28, 1968, Roald Fryxell flew to Washington, D.C., with the small gray travel case containing the reassembled skull fragments occupying the seat beside him. The next afternoon, flashbulbs popping and cameras cranking, "Marmes Man" made his media debut to the nation and the world at a televised press conference in the offices of Senator Warren G. Magnuson.

The Daugherty team's approach wasn't dictated by a longing for self-aggrandizement but simple scientific necessity. The Corps of Engineers was due to close the spillways of its new dam on the Snake by Christmas 1968; within days of that, the canyon of the Palouse would be drowned for fifty miles upstream of its mouth, the Marmes Rock Shelter along with it. Magnuson had already sponsored legislation to allow federal agencies to spend money to salvage archeological sites in danger of imminent destruction. The plight of Marmes was perfectly suited to putting some urgency behind that bill. Before Daugherty and Fryxell left D.C., they had lined up grants from the National Science Foundation, the Park Service, the U.S. Geological Survey, and the Corps of Engineers itself.

Salvage archeology was not an entirely new notion in 1968: American citizens had contributed generously to the international campaign to save the 3000-year-old Egyptian temple of Abu Simbel from the rising waters of the Aswan High Dam. But the crash program to investigate Marmes before it disappeared beneath the waves was the first major project of its kind in America, and even today it remains one of the most successful in terms of its results. Two more sets of skull remains of roughly the same age as Marmes Man I were discovered, along with hundreds of artifacts (including an exquisitely fine bone needle) indicative of the lifestyle of the ancient peoples who lived there. Also obtained was detailed information about the changing climatic conditions they had to cope with over the 10,000 years during which the shelter was used alternately as a burial ground, a living space, and, just possibly, a site for ritual cannibalism. And Jim Chatters, WSU anthropology freshman, was in the thick of the operation. "A good digger," Daugherty recalled, thirty years later.

Given the commitment of the Corps of Engineers to preserving the Snake River spring salmon run, there was no way to delay the opening of Lower Monument Dam. But so lush was the grazing at Marmes that the scientists were able to persuade their 1000-pound gorilla of a senatorial ally to browbeat a reluctant Corps into delaying the opening a few months to allow time to build a protective cofferdam around the site. But when the spillways were finally closed in February 1969, water rose inside the levee as fast as it did ouside: The Corps engineers had failed to notice that a thick band of coarse gravel underlay the site of their construction, allowing the rising waters a clear conduit beneath their $1.5 million folly. Thus, 120,000 square feet of the richest source of early human occupancy in the New World was lost, perhaps forever.

Apart from its incalculable impact on North American archeology, the Marmes miscalculation created a permanent state of tension and distrust between three generations of Army engineers and of scientists at the mercy of the Corps's passion for grandiose construction and waterways projects. Much of the time, the friction is concealed from public gaze, primarily because archeologists and anthropologists depend on the Corps and its insatiable appetite for digging for many of their opportunities to excavate. Furthermore, thanks to federal laws mandating historic-preservation studies, they are indebted to the federal government for most of the money to finance their respective sciences. But the friction is there, poisoning the relationships among officers and investigators who were still excavating in sandboxes when the waters closed over Marmes Rock Shelter. Jim Chatters had been an eyewitness to the catastrophe that generated the tension—a tension that, in the case of Kennewick Man, flared at last into open war.[2]

*T*hree years after Marmes Rock Shelter disappeared beneath the backwaters of Lower Monument Dam, Jim Chatters crossed the Cascade Mountains to work on his Master's at the University of Washington in Seattle. It was a very different kind of department from the one in which he'd earned his Bachelor's in 1971, partly because it was fashionable among faculty and students at the self-nominated "University of a Thousand Years" (est. 1893) to look down on Jim's alma mater, a mere "college" until 1959, and to deprecate (and envy?) Daugherty's gift for prying money out of Congress for his excavations and publicizing them to the hilt, once made. It was also a tense time in the UW department itself, which was still recovering from a purge of "old-fashioned" descriptive anthropologists, who saw their main job as getting acquainted with and coming to understand the cultures of living peoples, to make room for younger, more "scientific" faculty trained in the kind of system- and theory-dominated anthropology that had grown fashionable during the 1960s.

Jim's intelligence, diligence, and concentration earned the respect of his teachers. But other qualities were beginning to cost him sorely among his fellow students. Perhaps because they are professionally fated to spend a great deal of their time squatting in the dirt in lonely, distant places, archeologists tend to be a convivial, even rowdy bunch, and Jim, as his classmates remember him, lacked the gift for easy friendship. His impatience with ideas and opinions that differed from his own didn't help the situation, nor did his driving ambition, which made it difficult for him to see anyone achieve a higher station than himself. (One such person, exulting too openly in the notification that a prestigious journal had accepted his first professional paper, remembers being congratulated by Chatters with a raised middle finger.)

The net result of these behaviors was to make Chatters the frequent butt of his fellow students' horseplay. (Among a group discussing what each planned to wear for Halloween, one woman dropped to her knees and cried, "I'm going as Jim Chatters!") Perhaps this made Chatters even more of a loner, but it seems not to have reduced his determination to succeed at his trade. His Ph.D. thesis, submitted in 1982, is an exhaustive (and, at 452 pages, enormous) study of 4500 years of varying human use of a remote and isolated high valley in east-central Idaho.[3] The committee recommending the thesis's acceptance praised two aspects of Chatters's study in particular: his demonstration of how varying climate produces changes in human use of the products of the earth, and his determination to go beyond simple correlation in establishing cause and effect, suggesting instead concrete mechanisms for the changes observed. The recommendation ended with these words: "Chatters's dissertation may well serve as a model for the conduct of investigations relating settlement and subsistence parameters in environmental change."

Even allowing for the inflation customary in such documents, this recommendation would seem to presage a solid academic career. But archeology is a crowded field, with far fewer permanent teaching posts than applicants to fill them, and academic success is not the only—or even the single most important—avenue to employment. Apart from those vanishingly rare occasions when thesis research breaks genuinely new ground (or succeeds in calling some aspect of conventional wisdom into question), an aspiring archeologist needs a job (and funding) before she or he can hope to make a name in the field: to earn recommendations from respected seniors (or, even better, collaboration with one or more of them), build a good reputation among one's peers, and develop the ability to make an impression on strangers at the innumerable professional meetings where papers are presented, acquaintances made, and favor curried.

Temperamentally, Jim Chatters was ill equipped to play such games of professional advancement. For the first five years after his graduation, he held temporary teaching posts at Central Washington University in the small farming town of Ellensburg and served as associate director and coordinator of the local archeological survey, while earning much of his living performing the kind of unglamorous but increasingly frequent piecemeal survey-and-salvage operations mandated by federal law when development impinges on ancient settlement areas.

Chatters's fortunes seemed to change decisively for the better in 1987 with his appointment as a senior researcher in the Environmental Sciences Department of what is now the Pacific Northwest National Laboratory near Richland. By the time Jim came on board, the primary problem facing PNNL researchers was how to clean up fifty years' worth of plutonium and other radioactive waste accumulated here during the Cold War without precipitating environmental catastrophe. But like every facility covered by federal law, the former Hanford Nuclear Plant had to comply with regulations requiring that any terrain slated for development be surveyed in advance for "cultural resources" related to the nation's past.

Covering 560 square miles of lava and loess plateau along the last undammed stretch of the Columbia River, the Hanford Reach was rich in such resources. Before the Hanford Cultural Resources Lab was set up in 1987, thirty-one surveys had already been made in the area, turning up over a hundred sites of significant interest for students of the region's past, but the archeological potential of the area had barely been tapped. In Chatters's first year on the job, thirty-three more surveys were made, most just one step ahead of experimental nuclear waste-disposal projects, but enough areas remained inadequately studied to provide a scientist with many more years of fruitful research.

Unfortunately, Chatters's position at the PNNL was abruptly terminated in 1993, after an investigation involving employee time cards and work records. Officials of the lab say they are legally bound not to discuss the circumstances under which Chatters left its employ. Chatters himself has been less reticent, telling a number of people that he did nothing wrong beyond ignoring some typically finicking and shortsighted gov-

ernment accounting regulations, that he was offered the choice of re-
signing his post or being fired, and that he chose the latter because it allowed him to apply immediately for unemployment benefits.

Whatever the reasons for his departure, Chatters found himself
once again out of steady work and compelled to hustle against stiff
competition for such survey-and-salvage jobs as turned up round the
region. In some areas of the country, the Southwest in particular, rapid
population continues to drive rapid development, and salvage archeology is not only a booming but a lucrative business. On the Columbia
plateau, however, business for Chatters and his newly formed company
Applied Paleoscience wasn't so good: The Corps of Engineers' great
age of earth moving was long past, so most of the available jobs were
small ones: routine road building, housing and mall developments.
The bulk of the contracts available were on the lands of the many
Native American tribes that govern large areas of eastern Washington,
Oregon, and northern Idaho: the Yakama, Umatilla, Nez Perce, Colville, and Wanapum among them.

But such groups rarely have substantial funds to devote to such
work, and are not much interested in spending what money they have
on analysis and publication of results. They also typically lay down
many irksome limitations on the investigators regarding how remains
and artifacts are to be handled and what kinds of examination and even
description are suitable. At the age of forty-four, with one child of college age and another rapidly moving in that direction, Jim Chatters
found himself back in obscurity, a hired hand answerable to others, and
further than ever from the kind of professional distinction—and
fame—he had glimpsed as a youth.

Chatters did not give up. He took such jobs as offered themselves;
attended such conferences as he could afford to; wrote such professional papers as he could, no matter how skimpy the honorarium; and
volunteered his professional services to those, like Coroner Johnson of
Benton County, who couldn't afford to pay for them. No night school
class or Scout troop, no reporter with an archeological question ap-
pealed in vain. Should a break in the clouds appear, Chatters was prepared to take advantage of it. On the evening of July 28, 1996, as he

gazed at the contents of Bill Ashley's paint bucket, a shaft of light broke through.

*I*t is difficult to reconstruct the sequence of events between Johnson's delivery of the skull to Chatters's home and the press conference thirty days later announcing the discovery to the world: difficult but important, because much that happened in the weeks and months that followed happened because of actions taken—and not taken—during that short time.

The most striking thing about the Kennewick investigation, in fact, is the brisk pace of its unfolding and the way its course was dominated by the actions and choices of one man with no significant funding, affiliation, or support. Serious archeologists today are no longer pith-helmeted adventurers supervising a mob of sweating native day-laborers, but are, rather, chief executives of a diverse team of highly trained, opinionated specialists. A significant dig these days can extend over many years—even decades—and the publication of results can take longer yet. (The University of Kentucky's Tom Dillehay ended field work at Chile's Monte Verde site in 1983; the second and final volume analyzing the results of the dig ran over a thousand pages, included contributions from more than thirty specialists, and emerged in 1997.)

The only firsthand documentation of the Kennewick investigation is more modest: thirteen pages of desultory, telegraphic notes in Chatters's crabbed handwriting. As such they provide a fascinating but by no means complete behind-the-scenes record of that eventful month. In any effort to trace the roots of the two-year conflict among the interests of the federal government, academic science, and Native Americans, the events and findings recorded in the notes have proved less important than those *not* recorded at the time, which are still unfamiliar even to many intimately involved on one side or other of the affair.

The first diary entry, dated the day of discovery, July 28, 1996, shows Chatters very much in his forensic mode, making what seems to have been an easy call: "First impression looking at the calvarium and alveolar portion of maxilla is aged white male—this based on apparently dolichocephalic cranial index, postorbital constriction and obliteration of all but the squamous sutures—first impression 45 + [years of age]...."[4]

On the same page, Chatters's note of his first visit to the discovery site is equally laconic and unsensational: "Bone visible in failing light some 15 feet from cut bank in fine beach sand. Checked out bank and saw nothing in situ and no sign of b[each]pit, but light poor. Bank is strewn with historic artifacts, including blue willow ware ceramics, medicine bottle frags, square & round nails, other early 20th/later 19th century ceramics, sawn bovine and sheep bones...."

When he sat down the next day with an amateur colleague to begin his formal record of the remains, Chatters had a good deal of material spread out on the plastic sheet on the ping-pong table: substantial chunks of both arms and legs, a piece from the base of the spine, and one of the hinge-bones that connect skull and lower jaw, as well as half a dozen rib fragments. But it was the skull that arrested Chatters's attention from the first. And all but two of the fifteen notations devoted to it in this early pass emphasize over and over, in the sometimes arcane language of physical anthropology, the ethnic diagnosis of the night before: "Skull passes the pencil test, i.e.—caucasian like. Orbits are rounded, not square, again, caucasian-like.... Skull is dolichocephalic.... Distinct parietal bosses exist ... Mastoid processes are well developed and flat ... rather than round ordinarily seen on N[ative]-A[merican]skulls...." Also sharply distinct from Native American skulls to Chatter's eye (emphasis his): "Nose is long ... and projects very markedly.... There are *no* indications of wormian bones and *no Inca bone.* I'm certain of this."

Noting that few long bones of the limbs—the other primary source for assigning remains to ethnic categories—were yet available for measurement, Chatters made short work of them, concluding his first analytic session with two notes, the only indication so far of anything striking or unusual. The first—"Groove in carbonate cement in left femur fits a bone-handled steel knife found also on Sunday!"—is never mentioned again. The second—"Foreign object *inside* right innominate–iliac crest"—defined the entire future course of the investigation.

The record notes that Chatters and his helper Turner returned to the river bank on Monday afternoon and spent two hours searching the beach, turning up numerous additional fragments of human bone. They also record a visit to nearby Kennewick General Hospital, where Chatters asked a technician friend for a quick X ray of that "foreign object" embedded in a fragment of pelvic bone. "No luck," reads the final note for the date. "Just gray object, can't see form."

The notes reveal that Monday, July 29, was a very busy day. In fact, it was busier than the notes reveal: They make no mention of two phone calls—one to Chatters, one placed by him—each in its way important to the story. The first was from fellow archeologist Ray Tracy, who was employed by the Army Corps of Engineers district headquarters in Walla Walla, an hour's drive to the east of Richland. Someone in the office had noticed the *Tri-City Herald* item about human remains in Columbia Park, Tracy told Chatters. The last thing the Corps wanted was to interfere with the deputy coroner's investigation of a possible crime, but Columbia Park was federal property, administered by the Corps, so it would be wise for Chatters to apply formally to the Corps for an excavation permit under the terms of the Archeological Resources Protection Act of 1979 (ARPA). And maybe it would be a good idea to back-date the permit to the day of discovery, just to make sure there would be no question of correct procedure. Would six days be enough to tie things up?

Tracy did not, as he might well have done, ask Chatters why the hell he hadn't applied for an ARPA permit at his earliest opportunity rather than waiting to be asked to. He could hardly claim ignorance of the law; that same law had governed his six years of work at Hanford. Tracy remembers that he did, however, ask Chatters, in the jargon of the trade, "if he had indications of ethnicity. Chatters stated that the remains were not Native American. Rather, according to Chatters, they appeared to be the remains of a Euro-American, maybe a settler.... 'An old guy, maybe 60 or 70.'"[5] He did not mention the mysterious chip of stone in the subject's pelvis. If he had, Tracy might not so casually have given Chatters the name and number of the official who could issue a permit or have promised to put in a good word for the project.

The other phone call that does not appear in the written record is one that Chatters placed to Dr. D. Gentry Steele of Texas A & M University, a certified sage in the field of physical anthropology with a life-long interest in the first settlers of the North American continent. One does not waste Dr. Steele's time with questions about putative crime victims, old homesteaders, or even Native American remains much less than 10,000 years old. The origins and ethnicity of "Paleoindians," the earliest known inhabitants of the Americas, are his area of expertise. Nothing calls more into question Chatters's repeated assertions that he realized only gradually the extreme age of the remains than this call to the doyen of Paleoindian physical anthropology less than twenty-four hours after the first bone emerged from the Columbia mud.

The next day, Tuesday, July 30, the chief of the Corps's real estate division issued a permit to James C. Chatters, Ph. D., "to conduct work upon public lands owned or controlled by the Department of the Army," with the standard proviso that "cultural or historic materials collected under this permit will be turned over to the Walla Walla District, Corps of Engineers." With the assurance that his work was now properly authorized, Chatters spent Tuesday morning trying to see the gray object better, scraping away some of the sandy crust adhering to the pelvic bone and getting a look "at what is basalt or andesite object—flaked with serrated edges, rounded base—will CT scan this evening."

Before that, though, he had an appointment to keep a hundred miles north and west with Catherine MacMillan, a consulting anthropologist for the sheriff of neighboring Kittitas County, who had access to osteometric equipment and databases not available at home. Chatters showed MacMillan the bones he'd brought with him, saving the pelvic anomaly for last. "At every turn [she] would say 'white guy,'" his note reads. "Her observation, as mine was. 'If these bones were brought to me & the stone wasn't in the ilium, she'd have called it caucasian. She'll write a letter with her opinion."

With confirmation of his judgment of the ethnic affiliation of the remains in hand, Chatters drove back to Richland and another consultation: this one with Ken Reid, a Pullman-based freelance archeologist specializing in lithics—stone tools. Reid scrutinized the tiny patch of gray exposed in the pelvic bone, smaller than a little-fingernail in area, and agreed with Chatter's guess that the material might well be andesite or dacite, varieties of fine-grained eruptive rock used for weaponry in the region as far back as 13,000 years ago. Further than that he was not prepared to go.

Reid accompanied Chatters to Kennewick General to see whether a CAT scan could reveal more than the ambiguous X rays of the day before. Reid, the lithics specialist, gave a cautious reading of what he could see in the fuzzy electronic output of the CAT scanner. The material might be from a projectile, or it might not. The flaking and serrations Chatters had seen even before the CAT scan were not so obvious to him, but his caution failed to dampen Chatters's enthusiasm. That evening he made a rough sketch of the object he saw in the CAT scan (an ambiguous outline the size and shape of the side view of a cashew) and noted excitedly, "Appears to be a lance point with point end broken before entry. Really sliced its way in!"

Before returning to Pullman the next day, Reid spent a couple of hours at the find site with Chatters, getting badly sunburned while sloshing buckets of water, sand, and beach mud through a screen in search of more bone. They turned up a satisfactory trove, including more limb fragments, parts of a shoulderblade and collarbone, a kneecap, and lots more vertebrae and ribs. But after Reid's departure, Chatters went right back to work clarifying the ethnic affinities of the remains from the proportions and conformations of the skull, beginning "lightly to glue face and mand[ible]together for clear impression of form."

Another entry for the same day reveals that once again, Chatters's actions had been running ahead of his note taking. "Spoke with Floyd J[ohnson]about need to date skeleton. To resolve (hopefully) the ambiguity between the morphology and proj[ectile] point and some subtle, suspicious aspects like tooth morphology. He'll request 14C date. Which I have arranged with U[niversity of]C[alifornia]Riverside.... [D]rafted letter and sent to him to be typed and signed...."

The facility in question, R.E. Taylor's Radiocarbon Laboratory at UCR, is among the most prestigious in the field worldwide and has many more applicants for its services than it can accommodate. It is a tribute to Chatters's drive and powers of persuasion that in the less than seventy-two hours after the Kennewick remains were discovered, he had contacted the Taylor lab, had drummed up sufficient interest in their scientific potential to get Taylor colleague Donna Kirner to agreed to perform the tests not only immediately but free, and had arranged to ship a gram-scale sample (the fifth metacarpal, one of the smallest of the finger bones) off to Riverside.

The notes for July 31 also reveal that Chatters "[s]poke with Ray Tracy about decision to date and the reasons [for doing so. Tracy] concurred we were going about this correctly." This suggests that Chatters called Tracy to request permission to test. Tracy recalls that on the contrary, it was he who called Chatters (to request a progress report on the dig) and that Chatters mentioned the plan to test only in passing. But he confirms that he registered no disagreement at the time—a fateful decision, as it turned out, both for the Corps of Engineers and for Jim Chatters.

Radiocarbon dating has made remarkable progress since the days of Chatters's father, when testers needed large samples and had to wait months for enough of the radioactive carbon atoms they contained to decay for them to record a meaningful result. Subjected to the mass spectrometer method of dating developed in the early 1980s, a minute sample can yield precise results in a matter of mere hours or days. The only problem is that unlike the clumsy, old-fashioned method, the new procedure necessarily entails the destruction of the sample. When the sample is a fragment of elk bone or charcoal from an ancient campfire, this presents no problem. But when the sample is taken from human remains, science runs squarely afoul of ancient prejudice—and modern law.

In arranging for 14C testing of the remains, Chatters was pushing the envelope of what he was authorized to do under his ARPA permit. First among the twenty-two "conditions" in the permit issued Chatters is an injunction that insofar as possible, resources must be analyzed and results recorded in the field; that collecting material "solely for later laboratory analysis is discouraged"; and that the permit holder must specify in the application itself[6] "when laboratory analysis is indicated." It is understandable that Chatters didn't mention the possibility of such analysis in his initial application, when the odds were that the remains would turn out to be little more than a hundred years old. By the third of August, however, when his permit ran out and he applied to the Corps for a two-week extension, a bone sample had already been sent to the Riverside lab. But Chatters made no mention of waiting for results from Riverside as the reason for extending the permit. Instead, the application asserted that "low water levels" had impeded further investigation.

The clause concerning lab analysis was not the only one in the ARPA permit that might have given a cautious investigator pause. Condition "e" of the standard contract form reads, in part, "No Indian grave or burial ground may be investigated without permission of the governing council of Indians concerns [sic], which supplemental authority must be properly recorded with the official in charge of the designated area." There was as yet no evidence that the Columbia Park remains had anything to do with an "Indian" burial, but there was also no evidence to the contrary.

Furthermore, if the chip of rock found in the pelvis was, as Chatters believed, a stone lance- or spearpoint flaked and sharpened by hand, there was a strong possibility—if not a downright presumption—that the skeleton it was found in was "prehistoric" and hence was that of an individual deriving from the prehistoric inhabitants of the area, very likely one of the tribes among whom Chatters had been working for years. Getting permission from up to half a dozen tribes for destructive testing of even a tiny fragment of bone from a possible ancestor would surely have been a tortuous and time-consuming procedure—and by no means certain of success. By failing even to attempt to get such permission, Chatters was risking the charge that he had destroyed tribal property without permission and desecrated remains sacred under both traditional tribal law and his nation's.

With the mailing off of the bone fragment to Riverside, Chatters's "field notes" rapidly dwindle to brief and sporadic notations that would be uninformative to an outsider, but that nevertheless convey the sense of a great deal going on between the lines. On August 3, Chatters visited the beach and found a few more bone fragments. On August 5, he and two friends sluiced out more. On August 7, Chatters met with dentist friend Ken Lagergren to examine the teeth—with little result, to judge by his notes of the meeting. On August 8, Riverside's Kirner had good news: So well preserved was the bone sample sent to her that only half a gram would yield significant results. The next day Chatters was talking to UC Davis, where a specialist in extracting and typing DNA from fossil material thought she might be able to get results from the fragment of bone that Riverside didn't need, and at no charge.

The whole of the following week passes in three brief entries: On August 15, Coroner Johnson returned, and Chatters asked authorization for the DNA test. On August 18, Johnson issued his authorization, Chatters instructed Riverside to send its leftovers to Davis, and the Corps extended Chatters's ARPA permit another two weeks— again on the basis of changing water levels. Then the notes fall silent for more than a week, until the entry for Monday, August 26, in an even shakier hand than usual: "14C date is in—preliminary 8400!! Some trapper."

Injun Trouble 3

*I*t wasn't long after meeting Jim Chatters that Jeff Van Pelt had him figured for a hustler. Not that Van Pelt had a problem with that: He'd been hustling himself, long as he could remember. No, what graveled was the way that Jimmie always tried to dress his hustle up as Science.

The two men had got across each other often during the years when Chatters was the big, important, Ph.D.-toting archeologist in charge of the Cultural Resources Program at Hanford and Van Pelt was the self-described "little Indian boy" from the Umatilla Reservation across the river, coming up and telling the white folks at Hanford how to do their job. Chatters, in fact, had told people he was sure Van Pelt had had something to do with prompting the federal investigation into time sheet irregularities in his department that had led to the loss of his job.

Since then, Van Pelt had not had much occasion to think about Jim Chatters. He had, in fact, developed into quite the bureaucrat himself. As director of the Cultural Resoures Program of the Confederated Tribes of the Umatilla Reservation in Pendleton, Oregon, he was in charge of half a dozen full- and part-time professional archaeologists backed up by a highly trained crew of fifteen tribe members. In this capacity, he ran a summer cultural heritage program for tribal teens and an oral history project—this on top of his office's main reason for being: to supervise and enforce regulations governing the handling of known cultural sites along the Columbia, as well as the constant "inadvertent discoveries" made by dozer drivers and shovel jockeys turning the earth for the latest strip mall or housing project.

It was just such a discovery that brought Van Pelt and Chatters again face to face the morning after the fateful press conference at Kennewick City Hall. It wasn't "Kennewick Man" who was on the agenda, however, but a discovery made just the preceding weekend, when another casual beachwalker along the Columbia had stumbled across a jumbled pile of bones about a mile upstream from the site of the raceday discovery. History repeated itself, this time on fast-forward: The finder called the police; the police called Coroner Johnson; Johnson called Chatters, who showed up, looked the bones over, and without expressing any opinion this time about their age or origin took them home on his own authority, saying that his excavation permit from the Corps still had a week to run. The following Monday, he called the Corps to apprise them of his actions. (It was during the same call to his archeologist colleagues in Walla Walla that he dropped the bomb about the radiocarbon date he'd received just that morning from UC Riverside.)

Van Pelt knew about the Water Follies find, of course; for him, anything connected with the name Jim Chatters set the red lights flashing. But it was just one more inadvertent discovery in a year more than usually full of them, and by the time he reached the Corps to ask why Chatters and the coroner were dealing with an ancient skeleton from traditional Native American burial grounds, the Corps had already issued its permit to dig.

This later discovery was quite a different matter. Van Pelt already had a crew working nearby with developers who had stumbled over a Native American graveyard while laying out a golf course. The discovery area was a recorded occupancy and burial area. As Van Pelt saw it, the law was clear: Any investigation of remains from the site had to take place under Native American supervision, and Chatters's Corps permit to dig at a site more than a mile downriver no more gave him the right to investigate or even handle remains from another than it licensed him to dig in Egypt's Valley of the Kings. This just might be a chance to catch Jimmie with his scientific pants down.

The atmosphere at the discovery site that morning was tense, to say the least. Chatters was actively defending himself on two fronts, with the Indians (Van Pelt and a couple of Umatilla elders, Rex Buck and Bobby Tomanowash from the Wanapum tribe upstream, Alan Slickpoo from the Nez Perce) on one side, demanding to know what he thought he was doing taking charge of Native American remains without permission or even consultation, and a very upset Corps archeologist, Ray Tracy, on the other.

Tracy was in a spot. In picking up the Saturday discoveries and taking them home, Chatters was clearly out of line; on the other hand, Tracy was the guy who'd told the Corps to issue Chatters a permit in the first place, and since the press conference announcing the age of "Kennewick Man" the day before, the phone lines between Walla Walla and higher Corps commands all the way up to the Pentagon had been overheating. Someone was in deep trouble, and Tracy suspected that someone might be Tracy.

The colloquium in Columbia Park ended inconclusively and unsatisfactorily for all, with the tribal reps departing to draft a legal demand that the Corps force Chatters immediately to turn over all remains in his possession and Tracy heading back to Walla Walla for an emergency meeting called by the colonel in charge. For the 200-pound, six-foot-two Van Pelt, about the only redeeming aspect of the meeting was seeing the diminutive Chatters "squaring off like he was going to take me on. I mean he was mad: mouthing 'I'm going to get you Jeff Van Pelt, I'm gonna get you.' He couldn't take having some dumb Indian telling him what the law is, see? Indians aren't supposed to know anything about the law."

*T*he Umatilla Indian Reservation is only about an hour's drive to the southeast of Kennewick, but it somehow seems part of quite another world. From the front porch of Jeff Van Pelt's lonely house among the wheat fields a mile or two west of Interstate 84 and the flashing lights of the Umatilla's Wild Horse Casino, the soft, dark bulk of the Blue Mountains fills the horizon and the eye of the soul. The Umatilla and their allies the Cayuse and Walla Walla ranged far north of here across the lava plains of the Columbia to fish and to trade, but it was the Blues they called home, the Blues with their conifer forests, their high meadows rich in berries and edible roots, and their abundant game: deer, elk, bear, and wild turkey.

For Van Pelt, born on the reservation in 1955, home was working in the berry fields from five in the morning until six at night, turning over half his earnings to his parents when he got home, and hoping to hide some of the remainder before his older brothers got their hands on it. But even then, he had a plan. "I was a little miser. I remember once I had 97 cents saved up. My goal was a dollar. My mother took it to buy food. I didn't make that mistake again. I went down to the little country store and set up an account with the storekeeper. Nobody was going to take my money again.

"I was nine when my dad's drinking got so bad my mom took us kids with her to her people in the city—well, Corvallis. I still wanted to make some money. Somebody told me you could make money going around and pulling weeds for old ladies. I started in knocking on doors and saying, 'You got any work for me? I'll do anything!' And it worked: Cute little Indian kid, they'd give me 50 cents to do some chores, and pretty soon I got my own clientele, making pretty good money.

That wasn't all, though. Family would come over, they'd all be drinking, I'd say, 'I'll wash your car for a buck and half,' and they'd say OK. I'd wash the car but I'd also take the seats out and collect all the loose change, run all the bottles down to the store for the deposit. I made more money doing that than the dollar and half from washing the car."

Van Pelt's mother moved from the reservation to get away from his alcoholic father, but she didn't escape alcohol herself, and her new husband was not just a drunk but an abusive one. Van Pelt and his brothers finished their growing up pretty much on the streets, and life on the street, even in a jerkwater town like Corvallis, is a kind of education, assuming you survive it. Van Pelt survived. "I realize now that even when I was a kid I was spiritually strong. I just didn't know it yet. All during those years I went to a lot of different Christian churches on my own, sometimes taking the bus all the way across town, but without any of them pulling me in. I was always looking for something. I didn't find it, and by the time I was a young man I was in serious trouble—not legal trouble, but spiritual trouble. But those were the days when the Native American movement was really getting going, and I started attending meetings and going to conferences and learned there was another kind of spirituality than the white man's. And it was that finally brought me back to the reservation."

It also, circuitously, led him into his present job. Trying as an adult to look beyond the welfare-induced culture of dependency on the reservation, and past the culture of alcohol, gambling, and despair that stained his childhood recollections of the place, he discovered among older residents, particularly women, a connection to another kind of past. "There's enormous wisdom there for anyone who's willing to listen. I started finding out that the job of those of us who are on the earth right now is preserving what those who came before left to us for those who will be coming after us."

For Van Pelt the born hustler, the one realization almost immediately led to another: that preserving his tribe's heritage could have not only long-term spiritual but also immediate practical benefits. In 1989, an increasing number of federal programs were available to finance "heritage preservation." The Umatilla, for decades dependent on the smothering swaddle of the Bureau of Indian Affairs, had no system and no expertise for striking out on their own to join such programs, most of which could be tapped only through a complex grant application process for highly specific projects and goals.

Van Pelt had no expertise in such matters either, but if there was any action going, he had abundant faith in his ability to score a percentage. "I remember going to the tribal Board of Trustees and telling them that if they gave me the authority to negotiate contracts on behalf of the tribe, I could bring in a quarter-million dollars for education, training, and employment for tribal members over the next four years. Four out of the nine board members started laughing. And they said 'Jeff, we love your enthusiasm'—because I was being quite the salesman—'but come off it.' But I insisted, so they laughed, and said basically, 'OK, son, you just go down there to do that.' Well, I ended up bringing in almost *half* a million, through a pipeline project that was going through from Canada down to California. I was negotiating contracts; I was running field crews with a white archeologist from Boulder overseeing the work and writing up the reports. I think I was called something like 'Protection Officer.' I was making $21,000 a year, which for me was a lot of money."

So successful was Van Pelt's program that other members of the tribal administration coveted the new kid's power, and his department was absorbed by the better established Economic and Community Development Group. "But lucky for me they thought contracts just came in the mail to be signed," and after two seasons when the take fell to a low of $60,000 per annum, Van Pelt got his department back, no strings attached. "I knew I would. My whole life I had nothing so I had nothing to lose. Anything I can go into and come out with something I didn't have when I went in is a victory. So I never lose."

*A*s it turned out, the meeting that afternoon back at the Corps wasn't as bad as Tracy had feared. The main topic of discussion wasn't who was going to take the rap for the screw-up but rather how to minimize its repercussions. Despite the already extensive media attention given to the Kennewick Man find, the top item on the agenda was placating "tribal sensitivities and concerns" about the casual handling of the remains so far. In the words of a memo summarizing the meeting, written by the Walla Walla Corps's civilian press rep and spin doctor Duane "Dutch" Meier, "The District needs to make clear, unequivocal demonstration of its commitment to the tribes as being a compassionate and supportive partner in restoring the remains to a condition of proper interment with dignity and respect, and full compliance with the spirit and letter of all existing laws."

The laws referred to, of course, included ARPA, under the cover of which Chatters had conducted his wildcat investigation, but the decisive statute—the one under which the tribes were preparing to claim the Kennewick remains as their own—was a far more wide-ranging one. In 1990, after many years of agonizing negotiation between representatives of the scientific and museum communities and Native American groups, Congress passed and President Bush signed the Native American Graves Protection and Repatriation Act. NAGPRA proposed to do nothing less than begin to redress more than a century of casual appropriation, by American museums and scholarly institutions, of American Indian skeletons and artifacts.

For the Native Americans who had campaigned for decades for such a law, its passage was a spiritual as well as a legal triumph. Native American philosophical and religious ideas are as diverse as the hundreds of cultural traditions they derive from, but one in particular is widely if not universally held: that the dead remain connected to the living and to the physical remains they left behind. Disturbing such remains, even inadvertently, disturbs the moral fabric of the world, with incalculable but always negative consequences. "Repatriation," for them, is not merely a matter of restoring property to its rightful owners. It is an opportunity to heal the breach between living and dead by restoring the bones and possessions of the ancestors to the earth they had been torn from in the name of science, profit, or idle curiosity.

NAGPRA not only called for the return to the tribes of hundreds of thousands of items squirreled away for decades in museum vaults but also established guidelines to govern how future discoveries would be treated. Dangerously broad guidelines, as it turned out, but in the virtuous glow surrounding the act's passage, nobody thought much about exactly how the law would be applied or about what some of the terms employed in it—"lineal descendants" and "cultural affiliations"— might mean in practice.

On a quick general reading, NAGPRA seemed clear enough. Artifacts and human remains found on federal or Indian lands were presumptively the property of living Native American groups and were to be rendered up to said groups on demand. But one question remained open in an area like eastern Washington: Given the location of a discovery, which Native American group had the strongest claim?

In hindsight, it was incautious of the Corps conferees to assume that under NAGPRA, the only conceivable claimants to the ancient Kennewick skeleton were the five main tribal groups with historic ties to the lands along the mid-Columbia and Snake: Van Pelt's Umatilla, the nearby Wanapum, the more distant Yakama to the west, and, to the northeast, the Colville of Washington and the Nez Perce of northern Idaho. It was particularly incautious in view of the fact that at the time of the meeting, the chief legal counsel to the Walla Walla Corps, Linda Kirts, was enjoying a little compensatory time off after a series of twelve- and fourteen-hour days of litigation.

Nevertheless, before the meeting ended, a plan of action had been adopted, reflecting point by point the demands tribal representatives had made that morning as reported by Tracy: "Remains found at both sites should be reintered ... as soon as possible. No public disclosure of the discovery sites should be made.... The District will assume the costs related to reinterment of remains.... The District will also actively partner with the tribes to establish an acceptable procedure...."

Most significant of all, as it turned out, was the apparently innocent proviso dictating "minimal further media treatment of the discoveries. The District will initiate no news releases ... and will maintain a 'respond to query only' media posture.... PAO [Public Affairs Office, Meier's bailiwick] will make direct contact to its counterparts at the tribes before responding to media queries, to ensure their awareness of our sensitivity to their wishes for lowest-visibility handling of media attention."

While the Corps and tribes were maintaining an impassive united front to the outside world, activity behind the scenes was lively and continuous. As it happened, the Wednesday of the war council, Corps legal counsel Kirts was already back at work on Corps affairs but was out of the office, and on Thursday she was tied up in meetings all day. It wasn't until she arrived in the office Friday morning and her attention was called to the *Tri-City Herald* coverage of the Chatters press conference that she learned of the issue that had preoccupied her colleagues all week.

The recap of the story, as presented to her by staff archeologist John Leier, left her more confused than enlightened, but from the perspective of her own professional expertise, one issue came immediately to mind. There was no question that the land where the remains had been discovered was under the legal control of the Corps; it was federal property. The remains therefore were also federal property, to be handled and disposed of under the provisions of federal law. Why, Kirts asked her colleagues, had the bones been left in the hands of a county coroner or one of his minions in the first place?

The answers she got left her even more frustrated. It appeared that they did not even know in whose possession the bones were *now*. With only a few hours left before the country shut down for the long Labor Day weekend, Kirts picked up the phone to do some research of her own. Her first call was to Benton County prosecuting attorney Andy Miller, with whom she'd had professional contact in the past. No luck there; Miller was out and nobody knew when he'd be back. She might try calling the coroner's office, though. Kirts did so, with better luck. The bones? "Oh, yeah," said Johnson, "Jim's got 'em. At least, he's got 'em now. But in a week or so, he's taking 'em back to Washington so the Smithsonian can study 'em. Isn't that great?"

Kirts told Johnson that the bones would go to the Smithsonian over her smoldering ashes. If he wasn't familiar with the terms of NAGPRA, she would be happy to fax him a copy. Neither he nor Chatters had any business handling remains from U.S. property at all, let alone shipping them anywhere. As a matter of fact, he could consider this phone call a formal demand that he repossess the remains in question from his deputy and secure them under lock and key until a representative of the Corps could take possession. Similar notice would immediately be faxed to Prosecutor Miller, should there be any question whatever about the Corps's position on the matter. Clear? Goodbye, then, and have a nice weekend.

Kirts prevailed in the argument with Benton County officials, though at the cost of a coolness that remains to this day. Reached at home that same evening, Prosecutor Miller agreed to take control of the remains in Chatters's possession and directed the coroner to collect them and transfer them to the evidence vault at the sheriff's department.

Johnson did so, but not until Monday (by which time Chatters had shown the bones to one more colleague, Grover Krantz of Washington State University, and had extracted from him a promise to provide a letter stating his conclusions from the cursory examination). While a deputy sheriff waited in the car outside, Johnson watched Chatters transfer Zip-Loc'd bags of bones from the examination table in his basement to a wooden crate and secure its lid with duct tape. Then Johnson took the box to the county sheriff's "evidence locker"—a converted two-car concrete garage in the parking lot behind the county courthouse—and, under the watchful eye of evidence clerk Marsha Hart, stowed it in the wooden bin with his name on it.

The next day, county, Corps, and tribal representatives met for the first time and, despite a tense and acrimonious atmosphere, managed to agree that the remains should be placed in a secure repository on neutral turf until it could be determined who should take ultimate charge of them. The site selected was a secure storage locker in Building Sigma V of the Pacific Northwest National Laboratory near the Hanford Atomic Reservation north of Richland, and thence they were convoyed two days later by the cops, the Corps, and observers for both the Umatilla and the Yakama. At the time, no one suspected that the remains would still be in the same locker over two years later.

The cortège from the sheriff's lockup to PNNL was duly recorded in the pages of the *Tri-City Herald*. Unreported for the best part of a year and imperfectly reported even then, however, was another visit to the storeroom in Sigma V two days later, when, at the insistence of Jeff Van Pelt and others, the Corps allowed the box containing the remains—packed under the eye of the coroner and under lock and key ever since—to be opened.

For many of the Native Americans present, the primary purpose of the visit was spiritual: to pray over the remains and ask their owner's troubled spirit for patience and good will while his descendants did their best for him. For Van Pelt there was an additional, more practical purpose. So deeply did he suspect any statement or activity connected with the name Jim Chatters that "I wanted to make sure before we buried them that we were not burying nothing but a box of rocks." He is less certain why, if that was his own goal, he insisted in taking with him to the inspection ceremony one of the professional staff members working for him on the Kennewick Golf Course cemetery site.

Like most anthropology students, Julie Longenecker had received training in identifying and describing human bone, although that training, under George Gill at the University of Wyoming, was a good twenty years behind her. But when asked by her boss to take notes while each bone-containing bag was lifted gently and held up for her inspection, she found what she'd been taught coming back to her. She was able to ensure that most if not all the bones in the container were in fact human, and she made a three-notepad-page inventory of them: skull, jaw, vertebrae pelvis, limbs, digits. After the remains were re-packed and prayed over, Longenecker handed her notes to Van Pelt, went home, and pretty much forgot the whole affair. It was more than a year before they proved important after all: not as evidence for sci-ence, but as evidence of a crime.

The press, by contrast, was out in force, for the announcement late that Friday afternoon in Walla Walla, where District Commanding Officer Lieutenant Colonel Donald R. Curtis, Jr., revealed that the Corps was instituting formal repatriation procedures on the Ken-newick remains. The first of two required announcements of the fact would appear in the legal notices column of the *Herald* on Tuesday, September 17. Assuming that no further persons with a claim to the bones appeared—and who, apart from the five tribes concerned, might that be?—the mortal remains of "Kennewick Man" were likely to be back in the earth at an undisclosed location before Hallowe'en. Any questions?

*A*part from a passion for justice in the abstract, Colonel Curtis had good reason to push for a rapid resolution of the tribes' complaints. From coast to coast, the Corps of Engineers is involved in the construction, maintenance, and operation of literally hundreds of projects along the nation's rivers, and a large majority of those projects impinge in one way or another on "Indian lands." The possibility of costly and time-consuming litigation with tribal authorities over real or imagined violations of law, contract, or tradition is an ever-present hazard. There were more than enough opportunities for friction between tribes and engineers already. The last thing the Corps needed was a cause célèbre suggesting disregard for Native American rights or (more significant with a small but vocal minority of the white community) spiritual beliefs.

There was another reason for fast, firm action: As agents of the federal government, members of the Corps felt that they had been manipulated, betrayed, and made to look ineffectual and foolish by a few small-town politicians pursuing their own parochial interests. Such insolence could not be allowed to pass with impunity.

Fortunately, it seemed, the Corps was on firm legal ground in pursuing the Native American interests and its own. On September 18, a preservation officer at Washington HQ e-mailed the commander of the North Pacific Corps District, warning against precipitate action that could expose the Corps or the government in general to public and congressional censure. "I know that rapid resolution to this matter is essential," Paul Rubenstein wrote, "But the train has many passengers and I am concerned that Walla Walla's actions today may leave the DCW and HQ staff at the station." Only minutes later, NPD Commander Bartholomew Bohn replied, dismissing Rubenstein's concerns. "Legal consultations have already occurred between our counsel, Justice, and the federal court system. All risk to us seems to be associated with not repatriating the remains...." Then, in a sentence that no doubt came back often to haunt him in the months and years that followed, Bohn closed with these words: "Now that the locals understand the law and their potential liability, they are distancing themselves from [Chatters]."

Ninety-nine times out of a hundred, a policy of relentless, oblivious forward movement like that adopted by the Corps in this case achieves its objective, thanks to the media's dependence on fresh fuel to keep a story's fires burning. But the tactic works only when there are no parties with an interest in fanning the flames and a source of combustibles of their own. In the little more than two weeks since the putative age of the Kennewick remains had been announced, the Native American demands for repatriation had been thoroughly aired in local and national media. While the Corps crouched behind its wall of silence, reporters had been talking to anthropologists and archeologists about the case and seemed remarkably well briefed on potential weaknesses in the Corps's position.

The usual sensitivity of liberal media to matters touching on Native American spiritual concerns was absent from the questions being posed, and Colonel Curtis was not given an easy ride. If a professional examiner like Dr. Chatters claimed the remains found in Kennewick were "Caucasoid" and resembled "pre-modern Europeans," they asked, why did the the provisions of NAGPRA, intended to govern disposal only of Native American remains, apply at all? And apply or not, wasn't it necessary to submit the remains to further scientific study to determine exactly to whom they should be repatriated? Wasn't repatriation without examination merely an example of the progress of science being thwarted by religious superstition?

Colonel Curtis, square-jawed, brush-cut, and looking every inch a recruiting poster for America's New Action Army, did not deal effectively with these queries. Obviously uncomfortable about being in the spotlight, he just as obviously had not been briefed to respond to the questions he was being asked. On the evening news, his version of the Corps's decision to repatriate was reduced to an ineffectual sound byte: "I'm trying to balance scientific research desires versus protection of cultural and religious rights," he mumbled, visibly on the defensive. "Jeez, my organization consists of engineers and scientists. Of course it's tough, but the law I think is pretty clear on what I need to do."

The Corps' refusal to reconsider or delay its decision to repatriate with all deliberate speed did not play well in the media. The next issue of *Newsweek* questioned the decision, as did the *New York Times*. Dr. James C. Chatters, the distinguished archeologist whose intrepid sleuthing had revealed the immense antiquity of the Kennewick remains, was quoted opining that the remains were of unique, incalculable scientific import. And he was not the only one to say so; the director of the physical anthropology program of the Smithonian Institution itself agreed. What evidence did the Corps have to rebut testimony from such a distinguished quarter?

Shortly after the second legal notice of intention to repatriate was published in the *Herald* on September 24, Corps offices from Walla Walla to D.C. began receiving letters and faxes from scientists around the country condemning the decision to repatriate and demanding further examination of the Kennewick remains. Some of those protesting quoted newspapers and magazines as the source of their information. The majority though, in their similarities of phrasing and interpretation of the facts, betrayed a single shaping influence, though it was some time before harried Corps personnel realized what they'd been hit with.

Over the preceding weekend, Douglas Owsley of the Smithsonian and the University of Tennessee's Richard Jantz had forwarded an e-mail message to every member of the Society for American Archeology and the American Anthropological Association, stating (inaccurately) that the bones had been seen "only briefly" by competent examiners and were to be repatriated without measurement or study. Phone numbers and addresses were provided where recipients might "express ... displeasure."[1]

At the end of the same week, the commander's office learned that the scientists hadn't restricted their communication to their professional peers. Letters began arriving from the offices of several members of the Washington State congressional delegation, including both of the state's senators: freshman Patty Murray and Slade Gorton, chairman of the subpanel for Department of the Interior of the all-powerful Senate Appropriations Committee. All said they'd been receiving complaints from constituents protesting the Corps's high-handed and antiscientific decision to proceed so quickly and unilaterally toward repatriation. One of the constituents identified was archeologist James C. Chatters of Richland.

On October 4, "Doc" Hastings, congressional representative for the fourth district of Washington State, which includes the Tri-Cities, provided the press with copies of the letter he'd sent to the national commander of the Corps, Lieutenant General Joe Ballard. In it he stated his conviction that reburial of the skeleton without further study would be "a tragedy.... Because the bones are so extremely old and so little is known about this period in the settlement of North America, it seems wise to learn more about the skeleton's origins before arbitrarily determining custody on the basis of a single unsubstantiated claim of cultural relation."[2]

Despite his nickname, Representative Hastings is neither a medical nor an academic doctor, but as a Republican member of the powerful Rules Committee (subcommittee on Native American and Insular Affairs), his opinion carried weight. The pressure on General Ballard increased sharply one week later when a second letter from Hastings reached his desk, this one signed by Senator Gorton as well, chair not only of the Interior subcommittee of Appropriations but also of the Select Committee on Indian Affairs.

The legal theory advanced in the letter was already familiar from the questions being asked every day by unusually well-informed reporters. It had been developed in detail in a September 25 letter to the commander of the Corps's North Pacific Division in Portland from an attorney of that city named Alan L. Schneider. Beneath its sauce of legalese and a generous peppering of statutes cited, the argument boiled down to NAGPRA applying only to remains of Native Americans. Affiliation between tribe and remains must be established in order for repatriation to take place. Professional observers state that the Kennewick skeleton is "of a Caucasian or Caucasoid male, or that it represents an extinct population group that was not ancestral to present Native Americans." Q.E.D.: NAGPRA does not apply.

In his letter Schneider did not mention why he was volunteering his legal expertise, but the Corps's attorneys were not long in suspense about its origin. On October 16, a week before the deadline set for comment by the Corps, a lawsuit was filed in the U. S. District Court for Oregon by Portland, Oregon, attorney Alan L. Schneider on behalf of eight scientists of national repute, stating that they had been deprived of their constitutional rights by the United States of America, the Department of the Army, the U.S. Army Corps of Engineers in general, and three officers of said Corps in particular, and asking immediate declaratory and injunctive relief (and attorney fees) for the harm done by said deprivation of rights and for violations as well of NAGPRA and of the Federal Administrative Procedure Act 5 USC section 706, 42 USC 1981 and 1983.

New Laws and *4*
Old Science

*D*espite the expansive opinions
expressed in his September 25
letter of advice to the Corps, Attorney Schneider did not attempt a
frontal assault on NAGPRA. When
the parties to the lawsuit assembled
in the Portland chambers of the
Honorable John Jelderks the afternoon of October 23, 1996, the
points of law immediately in question were narrowly procedural. A
provision of the labyrinthine Administrative Procedure Act of 1983 allows any agency of the federal government, should it find "that justice
so requires, [to] postpone the effective date of action taken by it, pending judicial review." On its face, the statute seems only to allow a government agency to suspend its own procedures when convinced that
the action stipulated would produce manifest injustice. But in fact, argued administrative law specialist Paula Barran, when Congress said
"may" in writing the Act, what it really meant was "must."

Before that novel interpretation could be argued, though, another
point needed to be cleared up. Lawyers for the Corps, top litigators
from the U.S. Department of Justice, were asking that the suit be dismissed out of hand, grounding their request on one of the deepest-
rooted principles of common law: the concept of "standing." To be
able to bring a civil suit, plaintiffs must be able to demonstrate that

their own interests, not just those of other individuals or of humankind at large, are being compromised by the defendants.

To establish that such an injury faced her clients, Barran suggested that they were threatened with nothing less than loss of a sacred First Amendment right, the right "to study and work with an object of scientific or aesthetic interest." It was quite a reach to find a precedent in case law supporting this radical extension of the Bill of Rights, but Barran's researchers at Portland's high-powered law firm of Lane Powell Spears Lubersky managed to dig up a 1992 case in which a judge found in favor of a group called Defenders of Wildlife, challenging Interior Secretary Manuel Lujan's interpretation of endangered species law. "It is clear," the decision read, "... that the person who observes or works with a particular [thing] threatened by a federal decision is facing perceptible harm, since the very subject of his interest will no longer exist."

With such flimsy legal foundation to build on, the plaintiffs' main asset was their sheer eminence in science. Two of them, Douglas Owsley and Dennis J. Stanford (Owsley's boss as head of the Anthropology Division of the National Museum of Natural History) had the honor of appointments at the Smithsonian Institution. Geoanthropologist C. Vance Haynes of the University of Arizona and paleoanthropologist D. Gentry Steele of Texas A&M were among the undisputed "grand old men" of their professions. The remaining plaintiffs were less widely known in the broader field of anthropology, but in the small body of workers specializing in Paleo-Indian studies, they swung a good deal of weight. In his heyday in the 1970s, C. Loring Brace of the University of Michigan wrote several influential books on human evolution and prehistory.[1] The University of Wyoming's George W. Gill was a past president of the Anthropology Division of the American Academy of Forensic Sciences. With the Smithsonian's Owsley, Richard Jantz of the University of Tennessee was in the process of developing the world's largest computer database of ancient human materials in museum and university collections. Robson Bonnichsen, director of the Center for the Study of the First Americans at Oregon State University, plays a central role in Paleo-Indian studies as publisher of the quarterly newsletter on new discoveries about early occupation of the New

World, the *Mammoth Trumpet*. Most of these eight were present that day in Portland only in spirit, but all asserted, in affidavits submitted to the court, their conviction that "Kennewick Man" was likely to provide data of paramount importance to science and that his repatriation without further study would deal a serious blow to the discipine to which they had devoted their professional lives.

Perhaps as important as the plaintiffs' scientific eminence in convincing the judge to allow their case to proceed was the complete absence from the defendants' table of experts of comparable renown. Confident of their unassailable legal position, the government's attorneys chose to argue their case on a narrow, closely textual interpretation of the law. It was a decision they came in time deeply to regret, it allowed their opponents to portray the government's efforts to the wider world as mere legal obfuscation and quibbling, while at the same time laying sole claim to the scientific high ground.

Barran started hammering away at this theme in her very first arguments. Not merely her clients' right to study was at stake, she said, but the general public's right to know. The full force of the government's mighty bureaucracy was being brought to bear to deny all humankind's right to know its heritage, in order to cater to the religious beliefs of a few. Judge Jelderks did not let her arguments pass unquestioned, but he was clearly intrigued with their direction, and soon he asked the question all Barran's rhetorical skill had been employed to elicit.

Her chance came when Jelderks, musing on Barran's contorted arguments about the definition of the words "Native American" under NAGPRA, asked her, "What if you thought this was a skeleton of one of your ancestors, and the court took the position that it was governed by the Act in question. How would you challenge that?" "If it were one of mine?" countered Barran. "Yes," said the judge.

Barran pounced. "One of the speculations that a lot of us working on this case have had," she said, "is that it could in fact be one of ours.... The scientific evidence that we know to date is from those first three anthropologists [Chatters, MacMillan, and Krantz] who had a look at it.... They believe that the evidence says this was a wanderer who is not a lineal ancestor of any modern-day person in this country.

He doesn't belong to any of us.... [T]hese remains become the property of the American people...."[2]

*W*ith that high-minded assertion, the battle lines were drawn and combat began. It turned out to be no Blitzkrieg, though both sides originally expected quick victory. Rather, it was a prolonged, exhausting trench campaign, with progress measured in inches often as not offset by loss of ground in the next encounter.

Although in court the prospect of judgment receded ever farther as the case continued, in the world outside Judge Jelderks's chambers, the decision was in before the first snow fell. Thanks to the refusal of both Corps and tribes to take their case to the media, it was "science" alone that shaped public perception of the facts, interests, and issues of the case. In article after article, broadcast after broadcast, the same scenario was replayed, with commentary and color offered by the same small band of experts, always ready to provide a reporter on deadline with a pithy quote summing up the situation.

More often than not, the experts consulted were the scientists appearing as plaintiffs in the case. But as scientists, and highly reputable scientists at that, their pronouncements were taken by the press more at face value than those of the typical "interested party." The thrust of the plaintiffs' case, after all, depended (or seemed to) not on arcane scientific issues but on the broadest conceivable ground: their right to pursue truth wherever it might lead them, not in their own interest but the interest of all.

Reporters more inclined by experience to question what they are told found a kind of negative confirmation of the plaintiffs' position in the near silence of other scientists not involved in the case. An occasional dissenting voice was raised but rarely given prominence in reports of the progress of the case. Such objections tended to raise exactly the kind of technical points that seemed irrelevant to the main issue: scientific freedom versus religious particularism. They tended, also, to question the enormous scientific importance attached to the remains and to downplay the mystery and romance of their discovery.

No reporter, even the most jaded, welcomes information suggesting that the story that is worked on so long and hard is overblown, let alone a total waste of time.

In fact, there was substantial unease in anthropological and archeological circles about the direction the Kennewick Man story was taking in the media, especially when the plaintiffs in the court case presumed to speak for the profession in general. Unlike scientists in many other areas of study, however, anthropologists tend to keep their family quarrels out of sight. Even practitioners of such closely allied fields as geology and paleontology find the surface calm at meetings of anthropologists in eerie contrast to their own rowdier and more genially quarrelsome professional get-togethers.

True, the sanctimonious politeness of the public presentations is frequently offset by vicious gossip and character assassination behind the scenes. Even without resorting to such tactics, however, a case can be made that the scientists lending their names to the cause of *Bonnichsen et al.* v. *United States*, distinguished though they were, by no means represented a cross section of their profession and were ill-suited to represent its values and interests to the world at large.

For all its importance and the excitement it arouses in the public, Paleo-American studies is only a tiny subdiscipline of the whole sprawling field of anthropology. All of its practitioners could (and at conventions often do) fit comfortably inside a medium-sized assembly room. Most are on first-name (if not always friendly) terms, (not surprising, since most have known each other since undergraduate days.)

This is particularly true of the Bonnichsen Eight: all of them white, all of them male, all (with a couple of exceptions) pretty much the same age. No fewer than three—Steele, Jantz, and Gill—graduated from the same program at the University of Kansas within a year of each other, in 1970 and 1971. Steele proceeded to the University of Alberta, Canada, where he taught Bonnichsen. Jantz got an appointment at the University of Tennessee where he trained Owsley, who now works for Stanford. Only Vance Haynes, the most distinguished and respected of the plaintiffs, stands apart from the rest in education and interests.

Paleo-Indian studies by nature deal with very old material, but the eight plaintiffs also display a strong bias toward the study of human remains rather than human tools or the evidence of human occupation. Haynes, a geologist as much as an archeologist, has spent a great deal of his professional life developing a detailed picture of the changing lives of inhabitants of the American Southwest as they adapted over the millennia to the ever-changing climate of postglacial North America. Except for Bonnichsen, a digger and flint-chipper from his Idaho youth, the plaintiffs tend to be men of the lab and storeroom more than field workers or excavators. Bones are their business.

Physical anthropology has deeper roots as a serious science than the three sister disciplines traditionally associated with it in American academe: cultural anthropology, archeology, and linguistics. The Göttingen physician and anatomist Johann Friedrich Blumenbach may be said to have founded the discipline with his 1776 monograph *De generis humani varietate nativa liber* ("A book about the natural variety among human races"). The monumental illustrated catalog of human skulls Blumenbach published between 1790 and 1828 is what Lamarck was simultaneously attempting for other forms of life: to classify the varieties of humankind on a rigorous observational basis.

Blumenbach was a progressive in the context of his time: In the face of numerous theories—religious, philosophical, ethical, and scientific—that proclaimed the races of of humankind to be separate creations, he asserted the fundamental unity of the species, demonstrating with his skull data that size, shape, and detail varied continuously across the spectrum of peoples and societies. But he also recognized that in matters such as stature, body mass, skull shape, hair and eye color, and other, more subtle visible characteristics, the great mass of human beings tend to fall into distinguishable clumps. On the basis of his limited data, Blumenbach decided that five clumps were sufficient to define the whole range of variation. He called them Ethiop, Amerind, Malay, Mongol, and Caucasian.

Blumenbach didn't feel compelled to establish a ranking among his five varieties of humans. Advanced thinkers of the European Enlightenment from Voltaire to Chateaubriand declared the fundamental equality of all peoples of the world, and they often suggested, in their writings that a Europe ruled by absolute monarchs could learn a good deal about government and social justice from other cultures and traditions. Or possibly, as a cultured European himself, Blumenbach took for granted the caucasoid breed's superior capacities.

Later-born savants building their theoretical structures on the foundations laid by Blumenbach weren't so cautious, or so polite.[3] The great physiologist Paul Broca (1824–1880) is still honored today as a pioneer in the exploration of brain function. Only intellectual historians recall him in another role: as one of the founders of scientific racism. Arguing backward from conclusion to data, Broca pointed out that the most "advanced" race of humankind, the caucasian, also happened to be on average the most brachycephalic—from the Greek for "long-headed"—whereas peoples with rounder heads (Africans and American Indians, for example) hadn't in the whole course of their histories developed such evidence of superior mental acuity as the steam engine and the spinning jenny. The conclusion seemed obvious enough: Other things being equal, long heads good, round heads bad.

Another of the nineteenth century's greatest scientists, Louis Agassiz (1807–1873), lent his weight to such assertions and made others that went a long way further. Agassiz's groundbreaking discovery that the earth had experienced repeated scourings by great waves of polar ice was made in his native, homogeneously caucasoid Switzerland. His first real encounter with people of another race—the black slaves who staffed the hotels he stayed at on his first trip to America—shocked, revolted, and terrified him. The fearful rage expressed in his letters home has been suppressed until recently by Agassiz's biographers. But Harvard's new natural history professor was not shy about putting his fame and authority on the line in the racial debate convulsing the nation as the abolition movement grew in strength. In an 1850 contribution to the *Christian Examiner,* Agassiz unapologetically argued the inherent superiority of the white race and dismissed the unfortunate runners-up with an adjective each: proud Indian, submissive Negro, cunning Mongolian.[4]

Ten years after the publication of Agassiz's editorial, as the United States lurched toward civil war over the right of blacks to be considered the political equals of whites, the republic of ideas experienced its most far-reaching revolution since the publication of Newton's laws of motion. Ironically, in the field of anthropology, the initial impact of Darwin's theory of the evolution of species was to lend credence to the notion that the races of humankind differed not only in (literally) superficial ways but also in inherent capacity, potential, and worth. After all (said the champions of innate difference, in what passed for rational evolutionary argument at the time), evolution implies evolution toward something; the whole record of life on earth shows a progress from simplicity to complexity, from mindlessness to mind.

No one denies that race exists, that human beings aren't continuously variable but occur in groupings distinguishable not only by geography but also by appearance, language, culture, and technology. Is it likely that such radical differences don't reflect other, more subtle differences as well? But no need to argue the point on principle: Science would settle the matter. With enough data and the right methodology, race would become a parameter as neutral and convenient as stature or body weight. All that was needed was somehow to find a way to reduce the blurry compass of racial variability to solid yet infinitely manipulable numbers.

Europe gave America its most prestigious apologist for scientific racism, Agassiz. America returned the compliment with the work of Samuel George Morton, the Philadelphia physician whose 1839 *Crania Americana* became a kind of Bible for those seeking assurance that the abilities and characters of whole populations of human beings can be predicted by crunching a few simple numbers.

Morton himself accepted Blumenbach's racial categories (though he believed that each race derived from different roots), but the sheer size of his data sample and the obsessive meticulousness of his method kept his work influential long after his death in 1851. Skulls from all over the world were shipped to his storeroom, where Morton developed highly refined ways of measuring not only their shape and dimensions but also the volume of the braincase. So ample was the data and so clear its exposition that it was more than a hundred years before someone thought it worth the trouble to re-examine Morton's work.

That someone was Stephen Jay Gould, and his re-examination forms the centerpiece of his book, *The Mismeasure of Man*. Not only did Gould discover that Morton had applied to his data statistical techniques that skewed the results; the data itself was hopelessly corrupt. By modern standards of evidence, Morton's study was worthless. (This did not prevent supporters of the racial arguments of *The Bell Curve* from citing the *Crania* approvingly in the *National Review* and dismissing Gould's recalculation as a Communist plot.)

*F*ew anthropologists today—and certainly none of the plaintiffs in the Kennewick lawsuit—would acknowledge Blumenbach, let alone Broca, Agassiz, or Morton, as a direct intellectual influence. But their work laid the foundations of the science of physical anthropology as it is practiced today. Blumenbach's five categories of humankind have blurred around the edges, blended, and recoalesced in several different patterns, but the notion persists that racial categories are good for something, if only as convenient professional shorthand. The idea that one can define the relative intellectual capacities of the world's peoples by averaging the skull diameters of different populations is stone dead as a respectable premise for scientific investigation, but the idea that massaging the data derived from a sufficiently large and well-distributed sample of discrete measurements will produce some useful and enlightening information about the nature and history of our species remains.

A hundred years ago, the great cultural anthropologist and linguist Franz Boas was already decrying the crude evolutionary emphasis of many of his fellow scientists. So powerful was his preaching and so productive his example that cultural anthropology has been largely spared the ravages of relying on the application of overweening theory to inadequate data. But physical anthropology remains almost of necessity in the thrall of numbers. Its practitioners approach the study of humankind with calipers, microscope, and test tubes at the ready. Unlike cultural anthropologists, they rarely encounter the living correlatives of the material they study.

In consequence, many physical anthropologists were psychologically and politically unprepared to deal with the consequences for their profession of the rise of Native American activism in the late 1960s. The first efforts by tribal groups to reclaim the bones and gravegoods of their ancestors from museum and study collections were dismissed by the curators and scientists in charge as sentimental nonsense or deliberate mischief making. By the time it became clear that the claimants were serious and that legislators and judges were taking their claims seriously, it was too late to mount an effective countercampaign. In state after state, laws were passed recognizing Native American property rights in ancestral materials and setting up procedures for the restoration of at least some to the possession of the tribes.

When it began to seem that the U.S. Congress might pass a similar set of laws, many physical anthropologists reacted as though a loaded gun were being pointed at their heads. And in a way, one was: Suddenly threatened with the imminent loss of materials they'd taken for granted all their professional lives, the curators of the great North American assemblages of human skeletal and ethnological material—at the American Museum of Natural History in New York, at Chicago's Field Museum, at the Smithsonian, and in university collections such as those of the University of California—reacted with the ferocity of a mother tiger asked to turn over her kits.

The collections' strongest argument for retaining possession of their materials—that after decades in residence, vast portions still had not been cataloged, let alone studied—proved a slender support. If these bones were so mightily important, argued attorneys for the Native Americans, why hadn't they been studied already? Fighting a rearguard action all the way, physical anthropologists watched helplessly as Congress debated, passed, and sent off for signature a bill that seemed to threaten their very professional existence. And the experience was rendered even more bitter by the lack of support they received from colleagues in the cultural branch of anthropology.

A hearing before the board of trustees of the Nebraska Historical Society in 1988 is a particularly poignant example of the growing isolation of physical anthropologists within the profession as a whole. At issue was the question of repatriating the Society's collections of hundreds of Pawnee skeletons, collected over more than a century. Arguing stolidly for the overriding interest of science in preserving the collection for study, the Smithsonian's Douglas Owsley found himself in the curious position of claiming to be defending the interests of Pawnee yet unborn against those of living Pawnee sitting at the table with him.

Owsley did not prevail in Lincoln, nor did he and his fellows prevail in their testimony before Congress. But the fight continued on other fronts. Even after NAGPRA was passed and signed, many in the profession continued to mount open or guerrilla resistance. It took the better part of four years for administrative specialists in the various affected branches of government—primarily the armed services, the Forest and Park Services, and the Bureau of Land Management—to come up with procedures to implement the provisions of the law: In the meantime, there was plenty of opportunity for delay, appeal and study.

Inexorably, though, the law began to take its toll, and nowhere more so than in the storerooms of Owsley's own base of operations, the Smithsonian, where a task force to deal with repatriating the Native American contents of "the nation's attic" was created as an independent entity with its own staff and budget. This group was quite outside the authority of Owsley's boss Dennis Stanford, director of the anthropology programs for the National Museum of Natural History.

There was good reason for upper management at the Smithsonian to pay special attention to NAGPRA: Among the collections of the NMNH was one of the most notorious assemblages of Native American remains, collected at the behest of the U.S. Surgeon General during the U.S. Cavalry's campaigns against the warlike western tribes following the Civil War. Graves were dug up, corpses decapitated, skulls of particularly noted enemies sought out and shipped to Washington for study to support his theories of Indian inferiority and inability to adopt the benefits of civilization. After these remains had mouldered in the U.S. Army Medical Museum for decades, the Smithsonian had almost casually accepted delivery of the collection, never imagining what a weapon against its scientific credibility possession of such materials would one day be.

If physical anthropologists in general felt threatened by repatriation and NAGPRA, the mood among specialists in Paleo-America bordered on despair. With only a few dozen fragmentary sets of remains to study, and with only half a dozen or so of those securely dated, paleoanthropologists faced loss of the bulk of their entire database. Even more appalling: The late 1980s and early 1990s saw a sudden spate of important Paleo-Indian finds, more than doubling the effective size of the available material in less than a decade. But with NAGPRA in place, it seemed likely that such material would no sooner appear than it would vanish into the ground again, precious and irreplaceable information sacrificed to what seemed to many scientists nothing more than racist politics and religious superstition.

Two approaches to dealing with the crisis suggested themselves. On the defense side, scientists determined to study the precious remnants of early humankind in the Americas could argue that NAGPRA did not apply, and had never been meant to apply, to such ancient materials. Offensively, the campaign could be carried to the enemy by waiting for a case where it could be charged that NAGPRA itself was invalid, flawed, even unconstitutional. "Kennewick Man" seemed a perfect opportunity to test the law. At the very least, Congress might be prevailed on to clarify its intentions; with luck, a judge might be found who was prepared to toss NAGPRA out of court entirely.

In order for their case to be taken seriously, the plaintiffs had to establish two things in the judge's mind: the paramount importance to science of the discovery, and the remains' lack of resemblance to the physical traits of the peoples claiming relatedness to them under NAGPRA. After fifty years of exposure to televised litigation both fictional and real, most Americans are highly tolerant, if not utterly cynical, about the kind of shading of the facts considered permissible in the courtroom. Still, it's more than a little puzzling that the Department of Justice lawyers arguing for the defense failed to introduce countervailing evidence of any kind to put the assertions in the plaintiffs' affidavits in context, because both of the main points that the plaintiffs needed to establish were subject to question on factual as well as interpretive grounds.

First: Given the conditions under which the Kennewick remains were found, water-scattered and deprived of context, and given the limitations of the radiocarbon technique itself, a single date, while intriguing, has little or no scientific significance. Second: If in fact Kennewick Man was fully as old as he was announced to be, he was just one of half a dozen sets of Paleo-Indian remains found in the West over the last thirty years, some of which were certainly older, and most of which arguably provided more hard information for students of the ancient past of North America.

If those had been the only objections to be made to the plaintiffs' presentation of their case, the proper response might well have been a shrug: In a time so saturated with hype, what possible harm could be done by a premature announcement of results, accompanied by a little overstatement—particularly when no one conversant with the field would be misled for an instant? A third point, however, is not so easy to dismiss.

The terms "Caucasoid," "Mongoloid," and "Negroid" have deep roots in anthropological parlance and are sometimes still used by professionals, mainly forensic experts, in addressing peers familiar with their scientific meaning and limited application. But by the time James Chatters was a college student, and certainly when he was working on a Ph.D., such language was already so politically inflammatory that no responsible investigator cared to employ these terms in a public forum where they were likely to mislead the uninformed and certain to offend. As for the word "European," in a context purportedly as old as Kennewick Man's, it is not only inappropriate but meaningless.

After more than twenty years of working in the field, it is inconceivable that Chatters did not know how his words would be taken by the general public and, above all, by Native Americans. In many versions of the oral tradition revived among Native Americans since the Red Power upheavals of the 1960s, their people are autochthonous: born from the very earth they live on. Even for the vast majority of those of Native American descent who don't take such traditions literally, the notion that their ancestors were latecomers to the country or, even shared the land with others, is hard to interpret as anything but gratuitous provocation from a white extremist.

Perhaps their reaction to Chatters's statement was most cogently expressed by someone with no ethnic or professional axe of his own to grind: Professor Laurence G. Straus of the University of New Mexico's Department of Anthropology, a specialist in the Neanderthal culture of the European Mesolithic: "Instead of trying to forge a mutual interest with Native Americans in discovering who these ancestors of us all were, we throw it in their faces that they are not the original Americans after all, when that is the one thing they have in their cultures to hang on to in face of the overwhelming dominance of ours."

In the months and years of controversy and recrimination that followed the initial airing of his remarks, James Chatters and his defenders have often asserted that his words have been exaggerated, misconstrued, and taken out of context; that in any case his comments manifested no intention to offend anyone's religious or ethnic sensibilities; that on the contrary, the storm of protest ensuing on their utterance was fomented by "radical religious fundamentalists" determined to contradict the discoveries of objective science.

With the passage of time, all these defenses, plausible on their face and generally accepted by the media, have worn a little thin, while the motives of the "objective" scientists speaking out on the matter have begun to seem less pure and high-minded than they at first appeared. Far from being a naive blunder on the part of a plain-speaking scientist, Chatters's remarks (along with much of the plaintiffs' argument to the court in the case of *Bonnichsen et al.* v. *United States*) have come to seem more like an exquisitely calculated provocation, a sophisticated political maneuver masquerading as cool scientific method.

This is true even though the immediate consequences of Chatters's action and of the Portland litigation have been unmistakably destructive of their own interests as well as those of their opponents. The humane sciences, already badly hammered by cuts in research funding during the Reagan and Bush years and kept on the defensive by ideological hostility from the religious right, are likely only to suffer from the further politicization of the field due to the squalid fight over Kennewick Man.

The general educated public has suffered as well, because, ironically, the 1990s have been a period of remarkable progress in archeological studies. New discoveries both in the field and in the lab have contributed to a far richer and more nuanced picture of the human past than was available even as recently as a decade ago. The human past of the Americas in particular, still sketchy and ambiguous after more than a hundred years of serious investigation, is emerging at last into the kind of clarity that the ancient European and Near Eastern world has long enjoyed.

To appreciate what happened and what didn't happen in Kennewick, in 1996 and 9000 years ago, we need to go a roundabout route, tracing a million-year-old path across three continents. But in the honored epic tradition, it is best to begin the story *in media res* and to work backward and forward from there. Thus, let us start with a summer rainstorm a little over ninety years ago, falling on a corner of New Mexico so far north and east it's almost Oklahoma, where twentieth-century Americans first got a clear glimpse of the continent's deep human past.

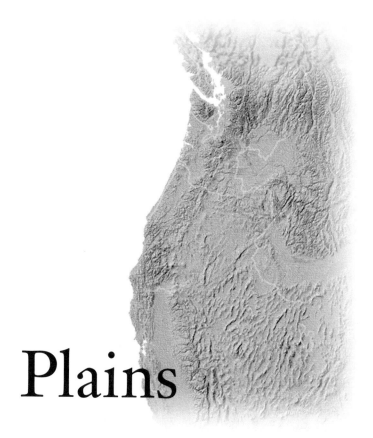

Plains

Clovis the Barbarian 5

*T*he scenery along Interstate 26 north of Albuquerque is impressive, but it's not seen to best advantage glimpsed intermittently from the driver's seat of a rental car between tandem tractor–trailers grinding by at 85 miles an hour. Only when the traffic abates, beyond the adobed-up suburban sprawl south of Santa Fe, does the view beyond the freeway's ribbon come into focus: To the south the endless rusty rim of Glorieta Mesa with its frosting of dark lava; the calm desolation of the flats between Las Vegas and Wagon Mound, with the Sangre de Cristos frosting the horizon to the west; the long run across a dinosaur-age sea floor, more and more pocked the farther north one goes by the rubble heaps of extinct volcanoes. Actually, these mounds of lava clinker are the youngest things in the landscape, but they add to the illusion that this broad highway is leading back to the beginning of all things.

"You want to allow four, four and a half hours altogether," the thready voice on the other end of the phone had said, "You take 26 all the way to Raton up on the Colorado border. At Raton you're going to take a right on state highway 72 and go up and over Johnson Mesa. It's beautiful up there, worth the trip all by itself. Coming down the other side of Johnson, start keeping an eye out for the ranch sign; turn in at the gate, go on past the buildings till you see some cars parked away down to the right by the dry riverbed. Park there and hike up the right-hand wash about a mile; you'll hear us working before you see us. You got a GPS? Maybe I better come down and meet you...." As it turned out, no cars were parked along the riverbank, and a mile, along the bottom of a twisty, thirty-foot-deep gully pretty much resembling all the other gullies in the neighborhood, gives one plenty of time to consider whether investing in a Global Positioning System receiver might have been worth it after all.

But sure enough, just before the arroyo narrowed to a deep, dark, brush-choked chink in the earth, it broadened on one side into a shallow amphitheater containing, at the moment: half a dozen scantily clad people sweating in a 10-foot-wide rectangular pit; a grid of stiff, foot-tall wires deployed over the sparse growth of the slope behind them, each topped with a diminutive colored-plastic banner; and, emerging from a white plastic tent big enough for the average wedding reception among the shrubby Mexican locust on the brow of the hill, a short, wiry, bearded man wearing a baseball cap and an expression of guarded welcome.

The excavation team working here under David Meltzer of Southern Methodist University in Dallas was not the first to have dug in this spot. Seventy years before the summer of 1997, this dry gully twelve miles west of Folsom, New Mexico, had been the site of perhaps the single most important discovery in the history of North American archeology, and Meltzer's team was here as much to throw light on the circumstances of that discovery as to expand on its findings.

"I originally got interested in Folsom for its historical import," Meltzer said, taking a not entirely unwelcome break from the back-wrenching work of driving augur holes down through twelve feet of stony topsoil to bedrock to chart the structure of the soils beneath the scrubby surface. "This is a really famous place." And for what may have been the third time that week and the tenth time that summer, Meltzer began methodically to explain the reasons for its fame, using the surrounding landscape like a picture book as he told the story of the Folsom site's discovery.[1]

"In August of 1908" (pointing toward the broad, flat lip of black lava looming less than a mile to the northwest), "thirteen inches of rain fell on Johnson Mesa, and it all drained this way, down the dry bed of the Cimarron River, blasted through the town of Folsom, and wiped it out. It also came down this tributary of the Cimarron, Wild Horse Arroyo, came through that very narrow gap you see across there at tremendous pressure, blasting into this side of the arroyo like a fire hose, cutting it much deeper than it had been before.

"A few days after the storm a cowboy on the Crowfoot Ranch named George McJunkin was out counting cattle and checking the fencelines along here—see the remnants of the fence across there, bits of barbed wire still stuck in the trees?—and he noticed some bones sticking out down at the bottom of the arroyo. McJunkin was real interested in everything around him, something of an amateur astronomer, naturalist, fossil hunter, and he noticed something a lot of cowboys might not necessarily notice, that these were buffalo bones, but fairly large ones, larger than he'd seen before, and he knew enough about bovine anatomy to make the judgment that this was something of potential interest.[2]

"McJunkin started telling people about what he'd seen, but the only person who showed much interest was a man named Carl Schwachheim, who was the blacksmith over in Raton, who was also something of an amateur naturalist. But, well, Raton, as you know having just driven from there, is thirty miles away, and though Schwachheim intended to get out here, it took a while—until December 10 of 1922, in fact. By this time McJunkin was dead and buried; last Sunday I took the crew over to see his grave outside of Folsom and look at the monument to the people who died in the 1908 flood.

"Anyway, Schwachheim and a friend, Fred Howarth, collected a sackful of the bones and tried to interest the state of New Mexico in doing something, but whoever they talked to elected to do nothing. Howarth, though, knew a rancher named Harold Cook, on honorary paleontologist at the Denver Museum, said, 'Sure, why don't you bring the bones to Denver and we'll take a look at 'em.'

"So Howarth did that—this was in March of 1926, so another four years had passed—and Cook and the director of the museum, Jesse Figgins, take a look at the bones and confirm that they *are* bones of extinct bison, 10% to 15% larger than modern bison, and that they want a complete skeleton for the Museum. Bear in mind the world was a much smaller place back then and the public could be thrilled by things that are mundane to us: If your museum could get a full-sized Pleistocene bison skeleton, people would flock to see it. So when they came out here to excavate in the summer of 1926, they weren't thinking archeology; they came to find some wonderful bison to put on display."

Though technically the Denver Museum was doing the excavating, practically speaking it was the Raton blacksmith Carl Schwachheim who got stuck with the job of (in his words) "polishing a pick" through the six feet of rock-hard clay above the bone bed. ("When I say hard I mean hard," Meltzer said. "I had to have a friend send me out a new augur last week because I pretzled my old one on this stuff: It's just a bear to work.") Starting in late May, it took Schwachheim a good month and a half just to get down to the bone level. It was already July 14 when he happened to notice a flake of stone lying in the loose dirt on the floor of his excavation: narrow, pointed, gracefully fluted along much of its length. It was an arrowhead or spearpoint, unmistakably the work of human hands.

Schwachheim may have been an amateur, but he knew immediately that the modest chip of rock at his feet could settle American archeology's longest-running and most acrimonious dispute. When the night train to Denver pulled into Raton, he was at the station, a letter to Figgins in his hand.

*A*lmost from the day they first reached American shores, European immigrants have been speculating on the origin of the peoples they were slaughtering, enslaving, and otherwise "civilizing." Thomas Jefferson, in this area as in so many others irritatingly ahead of his time, was one of the first to suggest[3] that "the Indians of America" might well have arrived in the New World from Asia via the Bering Strait—and at a very early period indeed. But for a century or more Jefferson's cool assessment was all but ignored in favor of more exciting stories a cast of characters including Welsh Druids, prehistoric Hindus, and the ten tribes of Israel misplaced during the Babylonian captivity of the Jews.[4]

At mid-century a number of scholars began trying to apply the test of evidence to the field, but it was not until the Smithsonian Institution's Bureau of Ethnology was established in 1879 that Native American history and folkways became the subject of serious scientific study. In the next fifty years, a great deal was learned about the evolution of Native American societies and culture before the arrival of whites, but one question of fundamental importance remained. In order to develop the enormous variety of languages and lifestyles recorded among the tribes, Native Americans must have occupied the Americas for a very long time. But how long?

In 1858, the year preceding the publication of Darwin's *On the Origin of Species*, a group of eminent British geologists had turned up stone tools unmistakably associated with the bones of animals extinct since the Ice Age. For a public still reeling from the announcement the year before of the discovery of the bones of an "apelike" being in a cave near Neanderthal on the lower Rhine, this revelation of the sheer depth of European history was both fascinating and disturbing.

No such unequivocal evidence of a deep human past had turned up in North America, but until now nobody much had been looking for any, and for the next half-century, hopeful amateur archeologists—and some not so amateur—periodically "discovered" human artifacts in sediments laid down not just at the end of the last Ice Age but long before it.

Between 1875 and 1925, in fact, the history of early-human studies in the New World can be summed up as a series of ever more far-fetched claims for antiquity, each in turn disputed and disproved by an increasingly skeptical scientific establishment centered on the federal government's U.S. Geological Survey and the Smithsonian's American Museum of Natural History.

The spearpoint that turned up that Bastille Day in Schwachheim's pit at Folsom was far and away the best sign yet that humankind was around when the extinct giant bison roamed the American plains. But there had been so many "false alarms" in the past fifty years that only absolute proof would satisfy the skeptics in Washington particular the skeptic-in-charge, Bohemia-born Dr. Aleš Hrdlička, the Smithsonian's curator of human remains and the thunderbolt-wielding enemy of anyone claim that human beings had been in North America before the retreat of the ice.

Schwachheim's telegram put Figgins in an awkward position. For one thing, a "floating" artifact (one not firmly, unquestionably embedded in its earthy matrix) would prove nothing at all to a stickler like Hrdlička: Who knows how long the arrowhead had lain there before it was noticed, how far up the excavation wall it had fallen from? But that wasn't Figgins's only problem, Meltzer said. There was also a certain "credibility gap" to contend with.

"See, for the last half-dozen years, Figgins and Cook have been writing up a series of sites which they say provide a clear indication that people have been here for half a million years, making pronouncements about issues and debates about which they know nothing. Figgins knows that with his track record, having a floating artifact isn't going to do him any good at all with the authorities. So he writes back to Schwachheim, 'Listen, if you find one of these in place, don't move it. Call me, and I'll come down.'"

But the rest of the summer passed with no further material found, and Figgins's self-restraint began to crumble. He wrote a report on Schwachheim's find for the journal *Natural History*, which duly published it the following spring. The unexpected result was a summons to Washington to show his "evidence" to Hrdlička, "who says, 'nice, very nice, but what you ought to do if you find anything is leave it in place and send out telegrams all around the country and tell people what you've got and they will send representatives to examine what you have.' Basically, see, Hrdlička doesn't trust Figgins for an instant, thinks 'I'm not going to believe anything this guy tells me he's seen. I want the pros called in.'

"In May Schwachhiem starts digging again. Same story, May and June slow going, nothing found; July, same, August, same. Then, on literally one of the last days of August 1927, he finds a point and this time it's in place. He writes to Figgins, who writes back instantly: 'Do not move it, do not touch it, don't let anybody near it, protect it twenty-four hours a day'—which is not a problem, because Schwach-heim's camping at the site—and out go the telegrams, one to the AMNH in New York, one to the Smithsonian, various other institutions, telling people, 'Arrowhead found in place....'"

This time Figgins was in luck. Hrdlička was in Alaska and didn't get the telegram, but a young colleague happened to be at a scientific conference in the town of Pecos. More important, the American Museum's respected paleontologist Barnum Brown was in Denver at the time, so Schwachheim's twenty-four-hour watch on his find could end after just five days, on September 4, 1927. "Nowadays if you want to know how old something is you take charcoal samples, send 'em to the lab, and wait six months, nine months, a year to get your dates back. In those days, resolving antiquity was relatively straightforward; you could do it in the field. All you had to do was establish contemporaneity: Were those points put down in the deposit at the same time that those animals were alive or dying? In this instance there was no question. You had a projectile point stuck in between the ribs of an extinct mammal."

The question that had troubled the field for decades was answered. Given the state of scientific art. There was no way, in 1927, to assign even an approximate calendar date to the Folsom remains, but it was clear that human beings had shared North America with creatures who roamed the plains not long after the retreat of the icecap that dominated the continent for many thousands of years. The big question was how long the sharing had been going on.

Apart from its intrinsic importance, the intellectual earthquake produced by a discovery like that at Folsom is followed by innumerable aftershocks. Investigators loath to hazard their reputations advocating ideas frowned on by the Establishment see a chance to break new ground, and thanks to Folsom, they knew where to look for new ground to break: Locate a streamcut through Ice Age sediments and look for big bones. When you find them, excavate around them looking for associated human artifacts. When you find *them*, call in a committee of experts to vouch for their antiquity.

In this monkey-see-monkey-do fashion, dozens of "Folsom-type" sites were found over the next decade or so. One, found five years later a few hundred miles south of Folsom, quickly came to overshadow its predecessor. Here, too, were human artifacts mingled with bones of prehistoric bison, but there were bones of horses also and, most exciting, bones of one of the biggest warm-blooded critters ever to roam the American plains, the mighty (and extinct) distant cousin of the African elephant, the mammoth. For the next half-century or more, scientific as well as popular imaginations were to be dominated by the story that seemed to emerge from the gravel and mud of Blackwater Draw: the epic tale of Clovis the Barbarian.

*I*t took nearly two decades for George McJunkin's discovery in Wild Horse Arroyo to bear scientific fruit. By the Roaring Twenties, everything was moving faster. Just five summers after Carl Schwachheim spotted his first in situ "Folsom point," Edgar Howard of Philadephia's Academy of Natural Sciences found himself fondling intriguing chips of stone and bone found in the neighborhood of the little town of Clovis, New Mexico, a few miles west of the Texas border.

Only months later, on November 12, 1932, more were found by a road crew mining gravel from a dry pond bed—and this time in situ. Before the end of 1933, Howard and his colleagues had unearthed not just spearpoints but a whole stone tool kit (flensing knifes, skin scrapers, awls, hammers), all in unmistakable association with the bones of slaughtered mammoth.

The spearpoints at Clovis were nowhere near so sophisticated in manufacture as those found at Folsom, but that was part of their fascination. By the 1930s, Darwin's theory of biological evolution, more or less crudely adapted, had deeply influenced the social and historical sciences, and though few leading archeologists would have put it quite so bluntly, in the prevailing mindset of the field, "less sophisticated" meant something very like "more primitive," and "more primitive" was a virtual synonym for "older." Folsom Man's reign as oldest-known inhabitant of North America had lasted barely five years.

Folsom's fame, beyond a narrow circle of professionals, was soon dimmed even further. Once discovered in the windblown dust of Blackwater Draw, Clovis artifacts were soon turning up everywhere: not just on the high plains of the west, but also out on the midwestern prairie, in the scrub pine flats of the south, among the deciduous forests of the eastern mountains, and on the treeless tundras of the far north. And everywhere they turned up (though not always at the same sites), the bones of mighty mammoth (or their shrub-eating relatives the mastodons) turned up too. Mexico, Central America, Colombia, Chile! Wherever excavators interested in American prehistory expanded their field of operation, it seemed that the Clovis folk had already been there and were waiting for them.

With widespread sites producing such a uniform range of prey, and with a uniform technology streamlined for dealing with it, it was probably inevitable that a very limited picture of the Clovis lifestyle would come to prevail. The reasoning, or what passed for it, ran something like this: Clovis people were obviously hunters—and pretty intrepid hunters, too, judging by their favored choice of prey. If the extinct mastodons and mammoths of America were anything like their surviving African relatives, they were formidable enemies: smart, mean, and, when threatened, aggressive in the extreme. No human culture in historic times has ever made elephants a dietary staple. The cost of killing one, measured in effort and risk to the hunters, isn't adequately offset by the potential payoff. (There's a lot of meat on a mammoth, but without access to a refrigerated van, most of the meat is going to rot before it can be eaten or even dried or processed into pemmican.)

On the other hand, Clovis fans said, the late Ice Age times when both mammoths and Clovis people ranged America were chilly enough to make natural cold storage feasible; and, in fact, caches of butchered mammoth meat have been found by excavators off and on over the years. But, the same people said, what's the point of arguing about why Clovis people specialized in killing mammoth, when it's obvious from the archeological record that that's what they did?

A hard point to argue, especially when the story is so attractive as it stands. Little wonder that no school text surveying American prehistory was complete without its dramatic watercolor of mighty Clovis hunters armed only with flimsy stone-tipped lances holding a rampaging mammoth at bay. Little wonder that every natural history museum worthy of the name featured a three-dimensional version of the same scene, sometimes with women and children cowering decoratively in the background.

The Great Depression was a rough time for Americans in general, but it wasn't so bad for archeology and its sister sciences, anthropology and ethnology. Public works projects—highways, dams, rural electrification—exposed traces of ancient occupation faster than scholars could document them. Activity slowed to a crawl during the four years of World War II, but exploded again thereafter, as the Corps of Engineers turned its mighty machines to peacetime development. But the basic picture of early human life in North America didn't really change, not even after 1947, when Willard Libby came up with the idea that was to earn him a Nobel Prize in 1960.

Libby had been part of the secret wartime Manhattan Project, which spent billions developing the atomic bomb, but his great contribution to human progress required little more than a garden-variety radiation detector (a "Geiger counter," in the parlance of the day), a desk calculator (a scratch pad would do in a pinch), and plenty of time. It depended on the recent discovery that every so often, a fast-moving neutron from the depths of space would hit the earth's atmosphere at just the right angle and energy to jolt an ordinary atom of nitrogen (the gas that composes four-fifths of earth's atmosphere) into an identity crisis, converting one of the seven protons in its nucleus into a neutron. Because the chemical nature of an atom depends only on the number of protons it contains, the jolted nitrogen atom, now with only six protons to its name, is nitrogen no longer but rather a somewhat overweight carbon atom, sporting two more neutrons than "normal" carbon, but indistinguishable for all practical purposes from the far more plentiful version.

On earth, those practical purposes involve the processes we call life. The carbon-14 produced high in the stratosphere quickly gets mixed with the rest of the atmosphere, there to be taken up by living things, usually plants, which get most of their carbon from the air around them. And there the story would end, except for one thing. The days of an eight-neutron carbon atom are numbered. Sooner or later, on a time table inherently unknowable, every atom of carbon-14 spits out an electron and resumes its original nitrogenous identity.

This doesn't happen very often—the odds that any given atom of carbon-14 will revert to nitrogen-14 within any given 5730-year period are exactly even—so carbon-14 decay is not much of a health threat to human beings (or to foraminifera, for that matter). But the process goes on just as inexorably after an organism is dead. Even though it is more than 5700 years before half of the carbon-14 you started with is gone, the remains of a critter that has been dead a long time are going to contain a good deal less carbon-14 than its living relatives, which are still frisking about inhaling their daily supply of the stuff.

The math involved in calculating just how much carbon-14 should still exist in dead organic matter after a given number of years is back-of-an-envelope stuff. Libby's genius was in realizing that if he could measure the residual radioactivity of old organic material of well-known age (wooden mummy cases from the Oriental Institute just a few blocks away from his office in the Chemistry Building of the University of Chicago, for example), he could use the results to set up a comparative scale to measure the age of material whose date wasn't known.

There were severe limitations on the technique: If the sample being measured was too recent, its radioactivity would hardly differ from that of the air around it; if it was too old, the "signal" would be too weak to depend on; and there was always a chance that the sample had been contaminated or otherwise modified. But the technique still revolutionized the field of prehistory.

Before radiocarbon techniques were devised, archeological sites could be dated only relatively, with Culture B asserted to be later than Culture A because its remains occurred in shallower layers of a given site or, just as often, because to the refined eye of an experienced archeologist, the style of its artifacts just looked later (more technologically advanced, more sophisticated in decoration, more just plain "cultured"). With radiocarbon, you could assign a number to your site, compare your numbers with other people's numbers, and establish priority in time on harder evidence than the style-sense of the excavator.

The first radiocarbon dates on early American material came along about 1950 and confirmed the feeling in the trade that Clovis sites were older than Folsom sites (a judgment also confirmed about the same time by further excavations at Clovis, where Folsom material turned up in shallower deposits). But as more and more dates became available, the whole field of Clovis studies got a major jolt. All the dates on Clovis sites discovered in 20-odd years of excavation were clustered in one 500-year period between roughly 11,500 and 11,000 years before the present—the archeological equivalent of an eye-blink. And before 11,500 B.P., nothing: no datable site of human occupation older than Clovis, anywhere on the continent. How do you explain a people whose traces suddenly appear across millions of square miles of tundra, grassland, forest, desert, and seacoast and just as abruptly vanish, all in half a millennium?

The sudden appearance was the easiest to account for. The reason why no evidence had been found of cultures earlier than Clovis, it was said, was that there had been no culture in North America before Clovis. And for good reason: Before about 12,000 years B.P., the great sheets of ice covering most of the continent north of the 49th parallel blocked any passage to North America from east or west. But the ice was already in retreat and, according to the Royal Canadian Geological Survey, was dividing into two separate masses as it retreated, one centered on the great shallow depression of Hudson's Bay, the other running northward along the coast ranges of Washington State and British Columbia. At some point—nobody was prepared to name a precise date, but 12,000 years B.P. was a nice round number—the two ice masses drew apart, leaving a narrow "ice-free corridor" that human beings could have traversed between the Alaskan interior (never covered by glaciers) and the great plains.

This 12,000 B.P. was a nice number not just because it preceded the known appearance of humankind in North America, but also because it fitted neatly into a sparse sequence of other numbers, these collected on the other side of the world. Early Russian excavators had been turning up evidence of mammoth hunters at sites in the Ukraine now known to date back to 35,000 B.P. By 20,000 B.P. similar folk were hunting near the shores of Siberia's Lake Baikal, thousands of miles farther east. More recently than 15,000 years ago, kill sites appear on the Aidan River, less than 500 miles from the Pacific coast. Not a lot of dots to establish a trend, but they connected nicely.

Those who were trying to trace the path of early humans into the New World particularly liked the trend line because it put people into far northeastern Asia just in time to take advantage of another symptom of the waning Ice Age. When ice caps rise on land, sea levels necessarily fall, and of all the world's ocean basins, the Bering Sea is affected the most when they do. Despite decades of study and analysis, the exact timing of the phenomenon isn't clear, but it is more or less certain that for thousands of years between about 25,000 and 15,000 B.P., the present-day Bering Sea was no sea at all but a vast plain, intersected by a myriad sluggish rivers, stretching more than a thousand miles north to south where today the oceans roll. If their timing was right—and it must have been right, because here they were—people could have simply strolled across the "land bridge" from Asia to the New World following their traditional prey, never knowing they were pioneering the peopling of a continent. This was important, because according to the conventional wisdom of the day, the boat, even in primitive form, had not yet been added to the species's technological tool kit. Once they were across the straits, it didn't really matter that ice may still have blocked the passage south. The Arctic plain of Alaska, never buried by the ice that weighed upon the rest of the northland, was available to keep the immigrants occupied until the ice-free corridor beckoned them to new hunting grounds.

Cultural anthropologists with field experience were dubious about the thesis. They had no problem with the route proposed, but they had a lot of questions about the timing. According to the Clovis-first theory, once they hit their stride, the immigrants spread ever outward and onward at a rate averaging fifty miles a year. No hunter–gatherer people known to science or history had ever behaved so, particularly when they didn't need to. And with dozens of never-before-exploited prey species deployed across their horizon, few hunter–gatherers have ever needed to do so *less* than Clovis.

However implausible the idea that the Clovis people spread from Canada to Mexico and from coast to coast in only about 250 years, said the believers, there was evidence that it happened. And far from weakening their argument, the rapid spread of Clovis culture actually reinforced it. Confronting a totally pristine, uninhabited continent filled with more protein on the hoof than humankind had ever seen before—not just mammoth and mastodon and bison but whole genera of huge and no doubt tasty critters ranging from ground sloths the size of an SUV to land turtles as big as a picnic table—Clovis hunters cut down their mostly vegetarian prey so efficiently and rapidly that they depleted the resource and were forced to move ever onward to areas where the pickings were richer. The disappearance of North America's so-called "megafauna" around the same time that humans arrived was no coincidence, the Clovisites said. In a 500-year frenzy of killing, Clovis hunters swept the continent clean of the big animals on whom their ancestors had preyed for millennia in Asia. Only when the proboscids (and the sloths and the turtles and the camels) were gone were these heroic hunters forced to exploit smaller prey and thus begin adapting to a less bountiful environment—a process that led, in time, to the development of the highly diverse cultures the European invaders found in residence on their arrival.

It makes a great story. It also sounds a little familiar. In fact, allowing for a somewhat expanded time scale and changing the skin color of the protagonists from white to red, the epic of Clovis the Barbarian could have been adapted wholesale from one of the most influential historical hypotheses ever propounded by an American: Frederick

Jackson Turner's theory of "the frontier." Put forward most persuasively in his 1920 *The Frontier in American History*, Turner's notion was that almost everything distinctive in the American character—from rootlessness to a fondness for firearms, from our attitude toward property to our choice of heroes—could be traced to the way the nation was populated. In a great wave of pioneers, adventures, and settlers rolling ever westward until they encountered the Pacific. The frontier has closed forever, Turner wrote, but the ideals, visions, and values it engendered live on to haunt American lives, which, compared to those glory days, inevitably seem a little diminished, even shriveled.

Historians still argue whether Turner's thesis holds water, but its influence on everything from political rhetoric to cowboy movies was incalculable, and by providing a richly evocative (if unconscious) template for thought, it continues to affect the way many anthropologists and archeologists think even today. Even before the 1960s, when the University of Arizona's Paul Martin put the finishing touches on what was to become the canonical version of the tale of Clovis's conquest of the New World, some investigators were trying to get people to consider evidence and theories that conflicted with its narrative. But so powerful was the appeal of Clovis's story that it has taken nearly twenty-five years just to begin to undermine its foundations.

How the West Was Won 6

*I*n the early days of American paleoanthropology, serious scientists faced enough problems determining where the first inhabitants of the New World came from and when they first reached their pristine hunting grounds to worry much about just what ethnic strain they might have derived from. In any case, the very year the American Revolution began, Blumenbach had asserted that indigenous Americans of all strains were skeletally distinct enough from the rest of human types to represent one of the five basic stocks of the species, and nothing that had been learned since indicated that the sage of Göttingen had been mistaken. Leave it to the amateurs and the loonies to speculate about the racial and cultural origins of the first Americans.

The amateurs and the loonies had hardly waited for permission. The shelves of nineteenth-century libraries groaned under the weight of massive volumes produced by religious zealots, weekend pothunters, and armchair lexicographers. Between them, they had unequivocally demonstrated the American Indian's derivation from Pharaonic Egypt, Old Testament Palestine, prehistoric Africa, and Homeric Atlantis, not to mention extraterrestrial and/or diabolic sources.[1]

Still, reports of human remains in what appeared to be very early contexts accumulated, bit by unsatisfactory bit: enough of them that in

1939, the pioneering plains archeologist Hanna Marie Wormington could publish a modest 80-page pamphlet, entitled *Ancient Man in North America*, pulling together all the reasonably reliable information about the first settlers of North America and drawing a few cautious conclusions about who they were, when they arrived, and where they might have come from.

By the time the Denver Museum of Natural History issued the fifth and last edition of Wormington's handbook in 1964, so much new data had accumulated that *Ancient Man* had expanded to over 300 pages. But one section had grown very little: the section dealing with early human remains, where the evidentiary line-up remained very much the same as it had been nearly twenty years before. All things considered, it was a sparse and unsatisfying data set. Apart from a human pelvis purportedly discovered together with bones of extinct megafauna by a Nachez, Mississippi, doctor in 1846, it consisted of

- a skeleton found in 1914 just off Los Angeles's Wilshire Boulevard in the famed LaBrea Tar Pits, also the last resting place of numerous Ice Age mammals;

- skull fragments discovered with mammoth and mastodon remains near Vero Beach on Florida's Atlantic coast in 1916;

- a skull found in an intriguing but highly ambiguous geologic setting deep in the gravels of San Francisquito Creek near Stanford University in 1922;

- half a dozen skeletons turned up on Angeles Mesa in southern California in 1924;

- three incomplete skeletons found between 1931 and 1938 at various gravelly sites in the moraine and lake country of west central Minnesota.

That, apart from a few fragmentary and questionable sightings in Wyoming and California, was that, until February 1954, when an amateur digger spotted human bones in a shallow, wind-sculpted gully in the sands near Midland, Texas, and called in the pros, among them paleo-Indian specialist Alex Krieger. As a result, Midland ended up the most carefully excavated and thoroughly documented human remains discovered to that date.

Considering their paucity and how widely dispersed the sites of their discovery, the finds enumerated by Wormington shared a striking number of characteristics. Skulls of Native Americans, both prehistoric and contemporary, tend to "brachicephaly"—round-headedness. To the extent distinguishable given their state of preservation, all of the skulls on Wormington's sparse roster (male and female, young and old) appeared instead to be markedly "dolichocephalic"—long-headed—the face and cranium being relatively narrow with respect to the distance from face to occiput.

Considering the enormous significance attached to Kennewick Man's supposed long-headedness by the scientist plaintiffs and the media alike, it is a little surprising that in 1964, Wormington found nothing particularly worthy of comment in the data. Indeed, her only reference to the matter appears as part of a warning against employing contemporary categories in describing "Paleo-Indians."[2] "It is an undesirable term if we give it a racial connotation. The later American Indians were Mongoloids, but this is not necessarily the racial type of the first comers to the New World. Some physical anthropologists think that the Mongoloid race represents a relatively recent development in Asia. Since we do not know when men first reached this Hemisphere, we are not in a position to say what their racial type may have been."

In the 35 years since Wormington's final update of *Ancient Man*, the number of well-documented Paleo-Indian specimens has grown substantially. Documentation on many of the most informative finds has appeared only since the mid-1980s: a 9800-year-old burial at Gordon Creek, Colorado (discovered in 1963, but documented only in 1971); the 10,000-year-old double burial at Horn Rock Shelter on the Brazos River northwest of Waco, Texas; another equally old burial at Whitewater Draw in the Sonoran desert southwest of Tucson, Arizona (1983); yet another the same year on Brushy Creek off Interstate 35 near Austin, Texas, perhaps a thousand years older still.

The majority of these finds and others made since exhibit the same general pattern noticed by Wormington. Where accurate measurements can be made, the general physical type tends to conform to the "Paleo-Indian norm," with long-headedness more common than not. Like Wormington, the authors who wrote up the finds leave it to the reader to draw any conclusions about ethnicity from the measurements provided or, in one case, go out of their way to warn against drawing conclusions. The authors describing Colorado's Gordon Creek site pointedly suggest that "archeologists and physical anthropologists ... leave the problems of morphology and 'racial mixture' until such time that sufficient data are available to justify a statistical analysis of human variability among early North American Indians."[3]

As the authors imply, "human variability" as used here is a euphemism for "race"—a term that, by 1971, no anthropologist would have used without long and serious consideration of the consequences. But their uncharacteristically blunt warning to the profession was necessary, for in the 1960s, just as race became a profoundly divisive issue in America, anthropologists and others intrigued by the peopling of the Americas were once more turning their attention to it. With the "when" question apparently settled to general satisfaction, the question of "who" for the first time seemed worthy of scientific consideration.

The impediments to finding an answer were enormous. Lack of data isn't the only problem one confronts when trying to extract useful information from early human remains. The old, old problem of distinguishing the relative effects of nature and nurture emerges here once more. The sparse remains of ancient North Americans available to physical anthropologists are scattered not only in space, but also in time: over more than 2000 years during which the environment they lived in was changing at a rate as rapid and in a manner as radical as our species had ever experienced, without much in the way of technology to buffer the impact. Such changes affect the amount, type, and quality of food available and the physical rigor of day-to-day existence, which in turn have effects great and small on the physiques of the persons experiencing them, whatever complement of genes they started out with.

On the other hand, certain distinct human traits are strongly or even totally defined by our genes. Hair and eye color are two of the simplest and most easily interpreted, but they are of little or no use to a paleoanthropologist. Some heritable characteristics have a longer shelf life, however, and among such characteristics, none are more long-lasting than teeth.

Today Arizona State physical anthropologist Christy Turner is probably best known to the public as originator of the theory that the Southwest's legendary Chaco civilization came to an end in an orgy of torture, witchcraft, and ritual cannibalism.[4] But his primary contribution to the field of American prehistory began nearly thirty years earlier with his publication, fresh out of graduate school at the University of Wisconsin, of a paper addressing the connection between tooth form and human evolution.[5]

Teeth, human and animal alike, are the paleontologist's greatest friend. Because of their very structure and function, they tend to survive far longer than any other part of the body. Many, if not most, of the specimens distinguishing the branches of humankind's slowly growing family tree are known only by the teeth they left behind. For students of the species' more recent past, they offer additional advantages. Evidence of decay and patterns of wear on teeth tell archeologists a great deal about the lives and diets of the people who once gnawed bones, cracked nuts, tenderized hide, and nipped thread with them.

For Turner, though, the most interesting things about teeth were those that have the least to do with their function. A great many characteristics of teeth appear to have virtually no effect on their usefulness to their owner. If you are living a dentist-free neolithic lifestyle, it is obviously a good thing to have teeth with thick, tough enamel, but it doesn't seem to matter much whether the teeth in question have two or three roots; whether their interior surfaces are smooth, grooved, or nubbly; or whether their outer edges flare out a bit or conform smoothly to the curve of the jaw.

It doesn't seem to matter much to Mother Nature, either. Like hair and eye color, many of the aspects of tooth form appear to result from fairly simple genetic mechanisms. Because such fine details of tooth shape don't noticeably affect the survival chances of their owners, there's no "selection pressure" nudging one style of tooth to be more prevalent than another. Best of all, from the point of view of a physical anthropologist interested in human origins, the fine details of tooth-crown shape and root conformation don't seem to have been major factors in mate selection throughout human history. Such characteristics evolve pretty much randomly, wherever the slow, random flux of mutation and chance encounters with the environment lead.

While still in grad school, Turner developed a way of describing dental variation in a fashion amenable to statistical analysis, and over the next couple of decades, he scoured the museums of the world for specimens. By 1990, he had in his database teeth from over 15,000 individuals, including most Paleo-Indian remains from both North and South America, a large group of precontact Aleut-Eskimos, and thousands of samples from all over the vast sprawl of the then Soviet Union. Turner left little to chance in his studies. Unlike most statistical analysts, he uses only samples he has seen and measured himself, in order to eliminate "interobserver error"—a more polite term for bias. Knowing that he might well never have another chance to see many of his samples a second time, he also tried to eliminate errors on his own part by viewing all of them (as far as possible) under the same intensity and angle of illumination, no matter what the location and date of the observation, and (as far as possible) recording no fewer than twenty-nine distinct "polymorphisms" for every sample, to allow the widest range of analytic possibilities back in the lab.

By the mid-1980s, Turner had run enough comparative calculations between individual dental "populations" to feel ready to define some terms and draw some broad conclusions. Ancient American teeth of all kinds, he announced, exhibit a strikingly narrow range of variation, suggesting that the people who once chewed with them all sprang from the same small, genetically uniform population.

Three traits in particular attracted his notice. Paleo-American incisors (the top and bottom pairs of biting teeth at the very front of the mouth) are statistically very likely to be "shoveled": more or less

deeply scoop-shaped on the inward side. In addition, two of the pre-
molars of the Paleo-American upper jaw (fourth from the front on
both sides) are likely to have just a single broad, blunt root instead of
the more common two. The first true molars on the lower jaw, on the
other hand, tend to have triple rather than double roots.

The only other part of the world where such traits predominate is
in northern Asia, China in particular. They don't occur, on the other
hand, westward and northward of China, among prehistoric populations of eastern Siberia, the Russian steppes, or Ukraine. If not from
the West, where did the ancestors of today's Chinese come from?
From down south, Turner's statistics suggest: from present-day southeast Asia, whence scion populations radiated in all directions beginning some 50,000 years ago, with one branch expanding northward up
the deep river valleys and along the coastline and at last westward into
Mongolia, south again into Kamchatka, Korea, and Japan, and across
the Bering Strait into the New World.

By the mid-1990s, thanks to advances in computer processing
power, Turner was prepared to refine his analyses much further, using
his tooth-type data to create "dendrograms" suggesting not just a
"most likely" family tree for humanity since the species's spread began,
but also (relatively) how far back in time each strain branched off from
the main stem.

Some of the results Turner has come up with seem counterintuitive, to say the least. Take, for example, his diagram charting the relative positions of all Old World populations sampled, which implies
that the Chinese–Mongol branch went off on its own even before the
divergence of African and European populations, that Australian abo-
rigines are near (geneticodental) cousins to the Danes, and that Nu-
bians and their neighbor Egyptians are about as far apart as the diagram allows.

But with other groupings, a pattern emerges that seems plausible as
well as enlightening. Turner's dendrogram concentrating on dental-
trait frequencies significantly represented in the New World shows a
very early split between a proto-European branch (destined to popu-
late early Europe and European Russia) and a proto-Asian branch.

Much later comes a split between a Chinese (and Mongol and southeast Siberian) branch and another, generically "Paleo-Indian" branch. The latter, judging by Turner's criteria, fans out into a northern group (California, the eastern United States, and Canada) and a blurrier southern complement, with overlapping subgroups centering in Peru, Central America, and Mexico.

Turner's studies have been widely influential and little criticized, partly because they are the only work of their kind available and he the only possessor of the raw data on which disagreement might be based. But the dearth of challenges to Turner's conclusions is largely due to their remarkable congruity with those generated by another set of studies in an utterly different branch of social science.

Almost half a century ago, Stanford University linguistics professor Joseph Greenberg put the finishing touches on the monumental study of the languages of Africa that he'd been working on for two decades. The applause from his colleagues had not yet died when Greenberg announced[6] that his next target would be the languages of the New World and that he could already see, behind their bewildering diversity, a pattern suggesting that all derived from just three distinct original tongues.

At the time, neither the profession nor the general public paid Greenberg's assertion a great deal of attention. The literature on New World languages was already enormous, but utterly disorganized: literally centuries of material sometimes casually, sometimes carefully, collected, for the most part by amateurs and enthusiasts with little sense of system and no knowledge of the principles of linguistic science. Nothing daunted, Greenberg kept beavering away at the ever-expanding corpus of evidence and, over thirty years after his initial declaration, finally published his *Language in the Americas.*[7]

Despite a hugely elaborate system of comparative etymologies (of common words, particularly pronouns), Greenberg in 1987 came to pretty much the same conclusion as he did in 1956. All the languages of the New World fall into three distinct groups: Eskimo-Aleut, spoken along the Arctic from the Aleutians to the east coast of Greenland; Na-Dene, dominant in the interior of Alaska and western Canada, as well as a large "island" in the American Southwest; and Amerind, spoken essentially everywhere else from Puget Sound and Newfoundland to Tierra del Fuego.

Greenberg's three-way split on the language front came in remarkably handy for theorists who had been trying to build a plausible picture of the settlement for the New World from the pitiful and scattered shreds of archeological material. It was powerfully reinforced by Turner's tooth data, published around the same time. Suitably tweaked, both dental and linguistic data, combined with present-day geographic distributions, suggested a clear temporal sequence, with three genetically and culturally distinct "waves" of settlers crossing Beringia to the New World at widely separated periods following the retreat of the ice.

First, according to the new synthesis, came the ancestors of Wormington's Paleo-Indians (the Amerind strain in Greenberg's terminology) entering North America at the usually postulated time of 10,000 to 12,000 years ago. These peoples spread so far and so fast that the Na-Dene, arriving in a second "wave" 2000 to 3000 years later, found the continent already well occupied. This explains, for believers in the three-wave theory, why their penetration was limited primarily to peripheral areas such as the northern tundras and the Canadian northwest coast, with only one breakout thrust into what is today Apache and Navaho country in the U.S. Southwest. The Eskimo-Aleut arrived last, only about 2000 years ago, and with their highly specialized boreal lifestyle were able in only a few hundred years to dominate the New World Arctic.

Combined, Greenberg and Turner's evidence yielded a clear, plausible account of the populating process—an account easy to grasp and communicate in lay terms and, better yet, consistent with what was known and believed already. In just a few years, the "three-wave" theory came to dominate discussion of the population question by both anthropologists and archeologists at professional conferences and in the popular press.

Indeed, most of the support for Greenberg's hypothesis came from outside his own profession—from archeologists and anthropologists with no special training in the arcane lore of linguistic analysis. Members of his own profession had varied reactions to Greenberg's magnum opus, ranging from enthusiastic support (for the most part from former students and members of his own department) to guarded neutrality and, in many cases, heated disagreement.

Like the other humane sciences, linguistics in recent years has drawn heavily and often productively on mathematical and statistical methods in analyzing its data. But because the bulk of its material is collected under uncontrolled conditions, under conditions that often cannot be replicated, and according to no single, universally agreed-upon method, intuition and insight still play a large role in any attempt at synthesis. Greenberg, said his critics, had depended far too much on subjective judgment in his analysis, seeing unmistakable affinities between tongues widely scattered across the hemisphere, where others saw only coincidences of sound or form. They also marshaled their own tables and charts and statistics showing that Greenberg had a tendency to downplay data sets when their numbers and patterns didn't bear out his thesis.

Fairly typical of Greenberg's antagonists were two specialists in indigenous American languages, Ives Goddard and Lyle Campbell, who summed up a vigorous attack on Greenberg and his allies in a mid-1990s paper as follows: "[R]eliable knowledge of the linguistic history of the American Indians is currently so incomplete, for all but the shallowest levels, that it is compatible with a wide range of possible scenarios for the peopling of the Americas.... We are particularly concerned that the classification presented by Greenberg should not be accepted as a reasonable working hypothesis simply because there is nothing else with the same far-reaching scope." As Campbell and Goddard saw it, Greenberg's very method was as flawed as his data was undependable: "Greenberg's insistence that hypotheses of classification validly precede hypotheses of history has produced an indiscriminate mass of unverifiable conclusions."[8]

Nobody has taken on Turner's arguments from dental evidence as vigorously as Greenberg's opponents have assailed the validity of his conclusions, but behind the firm scientific façade of numbers, charts, and diagrams, the dental data are vulnerable to some of the same strictures. The central problem with analyses like Turner's is that they too are based on statistical methods, with results that depend not only on the data poured into them, but also on the experimenter's choice of formula to crunch them. By the time Turner began his large-scale studies of tooth conformation, Greenberg's three-wave thesis had been around for decades, and it may have influenced Turner's choice of the techniques and weightings he applied to his data.

And no matter how sophisticated the mathematics that go into them, statistical methods by their very nature can never produce proof of truth; they yield only a probability that some assertion or other is true—and often a probability impossible to estimate or demonstrate by any other method. The science of statistics was founded by a gambler, and no matter how disinterested the motives of those who use it, they must constantly answer one question: "Here's your result—how much are you willing to bet that it's true?"

Another, less philosophical question faces all who would like to draw on the power of statistical methods in making sense of the variety of existence: "How hard are you willing to work to master them?" Number crunching has been an option for the humane sciences ever since Charles Darwin's brilliant cousin Francis Galton began diagramming the variability of every human trait from grip strength to musical genius in the last half of the nineteenth century. And there has been no shortage of economists, sociologists, education theorists, and other believers in progress through science eager to apply the latest formula developed by mathematically inclined colleagues to their data.

But statistical thinking remains difficult for most of us, including many otherwise competent scientists. An intelligent evaluation of the techniques and results in Turner's work is beyond the capacities of many anthropologists, especially those trained before a course in advanced statistics became a standard element in graduate school. They must, like the rest of us, take such results on faith, trusting to others more deeply indoctrinated to raise an alarm should the methodology or results seem dubious. But the only qualified people likely to take on the thankless job of evaluating someone else's experiment are those with their own theoretical axes to grind, so how much does their disagreement really tell us?

If the proliferation of subtle statistical methods left many an anthropologist feeling like a fourth-grade arithmetic whiz plunked down into an advanced calculus class, the *latest* set of methods to enrich the discipline might as well be quantum theory. Since around 1980, molecular biology has completely revolutionized understanding of our species's biological past, but the revolutionaries have for the most part themselves been molecular biologists, speaking a dialect of scientific discourse that is virtually incomprehensible to old-style social scientists of all stripes. Like us, the latter can struggle through an argument about tooth varieties or skull conformations without understanding much about the mathematical techniques that led researchers to the conclusions they reached. But population genetics does not yield so easily to superficial study. Worse yet, this most promising tool yet in deciphering the evidence of our species's spread across the planet routinely produces newsworthy, even sensational, results—but through means that not one in 10,000 of us is competent to judge.

Biology by the Numbers 7

Sometimes events in the real world contrive to come together so preposterously as to make the eerie tales from Rod Serling's "Twilight Zone" seem pedestrian. Consider the strange case of one Adrian Targett, married, childless, 42 years old, for two decades a history teacher at the Kings of Wessex School in the small, quiet, southeast English town of Cheddar. Apart from the cheese that has made Cheddar's name a household word, history is about all the town has to offer the outside world. But it is peculiarly rich in that resource. Located at the mouth of the picturesque Cheddar Gorge, which cleaves the limestone escarpment of the Mendip Hills, Cheddar lies on the northern edge of the great Somerset Plain, rich in evidence of human occupation since the peak of the last great European Ice Age, dense with legends of Arthur, Merlin, Excalibur, and the Grail.

More prosaically, Cheddar is also home to the mortal remains of Cheddar Man, found in a cave just inside the nearby Gorge in 1903 and dated decades later to about 9000 years B.P. A few more decades along, members of the Institute of Molecular Medicine at Oxford University began wondering whether the cool, damp cave environment where Cheddar Man had whiled away the millennia might have favored preservation of some of his genetic material. And, indeed, enough DNA was found in a root of one of Cheddar Man's molars to produce a genetic profile.

This research attracted the attention of a West Country television station, which went out looking for local residents willing to contribute their own DNA samples to ascertain whether Cheddar Man had any surviving relatives. Five of Cheddar's oldest inhabitants were persuaded to allow researchers to dab inside their cheeks with a cotton swab to sop up some sloughed off mucosal cells. Another fifteen volunteers were found among the pupils at Kings of Wessex School, after the accommodating Mr. Targett had submitted to the procedure himself to show it didn't hurt.

And it was the bemused Mr. Targett who found himself a nine-days-world-wonder on March 7, 1997, when the head of Oxford's Cellular Science Department, Bryan Sykes, announced that Targett was a linear descendant of Cheddar Man and, as holder of the World's Longest Pedigree, a shoo-in for inclusion in the next edition of the *Guinness Book of World Records*.

There's something inherently appealing about a story that pairs a mild-mannered provincial English history professor and a 9000-year-old cave-dwelling murder victim, and it would be churlish to criticize the media for making hay on the occasion. (Even Targett's wife Catherine got into the act, joking to the cameras, "Maybe this explains why he likes his steaks rare.") But the real science news produced by the study rather got lost in the shuffle.

For most of the twentieth century, European prehistorians have disagreed about who, exactly, modern Europeans are. Are they for the most part descendants of the same people who painted the caves and left behind the shell middens? Or are they relative latecomers, part of a wave of farmers moving west from the birthplace of agriculture in the fertile crescent of the Middle East, replacing the sparse early hunter–gatherer peoples of the region or drowning their genetic trace through sheer numbers? Targett's being related to an inhabitant of the westernmost extreme of the continent provides strong support for the first hypothesis. Farming spread westward across Europe, it seems because residents already there were smart enough to learn from their neighbors to the east, not because they sat passively by while someone else took over.

Reporters, however, weren't interested in the niceties of European prehistory: They wanted to talk genealogy, and Oxford's Sykes gave them what they wanted, though in a form where the more knowing could read between the lines. Targett and Cheddar Man "shared a common ancestor about 10,000 years ago, so they are related," Sykes said. He did not add just *how* nearly related; that would have spoiled the story rather. Because although Mr. Targett was indubitably the nearest living relative of Cheddar Man among the twenty or so people tested, the odds were that, given the current population of the British Isles (50,000,000 in England alone), Man almost certainly has many closer living relations, all blithely unconscious as their read their Sunday tabloid that only cruel chance had deprived them of a brush with weekend stardom.

The method used to trace Cheddar Man's relations, mitochondrial DNA analysis, was the first of several techniques in molecular biology put to work to illuminate the past. Thanks to one finding in particular, it is also the best known by far to the general public. Few interested in the history of our species have failed to hear of the lady known as "African Eve." So often and so confidently is this genetic matriarch cited in popular articles that many of us know more about her than about our own great-great-grandparents: where she lived and when, and all the generations that sprang from her womb. But unfortunately, Eve is neither so definite nor so uncontroversial as she seems.

Almost as soon as England's Frederick Sanger found a way to sequence DNA quickly and efficiently, researchers realized that useful knowledge about how DNA controls the processes of life might accrue more quickly through the study of simple systems: viruses, for example, and a virus-scale subunit of all animal cells called the mitochondrion.

Nobody knows exactly where mitochondria come from. Like viruses, they replicate themselves from their own little string of DNA, but unlike viruses, they can't exist outside a living cell. The most prevalent theory is that four billion or more years ago, some viruslike entity found a way to collaborate with an early one-celled life form instead of treating it as prey. However the collaboration began, it's too late to get out of it now: Mitochondria have become stripped-down, single-purpose machines for converting "food"—oxygen, glucose, and fats—into adenosine triphosphate, the only fuel the cell can use to drive its innumerable processes.

For a molecular biologist, mitochondria have two characteristics that make them terrific little subjects for evolutionary studies. First, the DNA containing their thirty-odd genes—mtDNA for short— evolves from five to ten times as quickly as the DNA in the nucleus of the cell the mitochondria inhabit. That's still not awfully fast; in a year, a molecule of mtDNA suffers on the average one mutation per billion possible sites. But it's fast enough so that over the roughly five million years since our ancestors parted genetic company with the chimpanzees, a lot of difference has been able to accumulate.

The other characteristic comes in even handier. Because mitochondria live and work and reproduce floating free in the cellular medium, they play no active part when cells divide. Half the mitochondria in a dividing cell, more or less, end up in one "daughter" cell, the remainder in the other. In one kind of cell division, however, the divvying up doesn't work: sexual division. Egg cells get a full complement of mitochondria; sperm cells don't get any. Thus whether the result of a union between egg and sperm turns out male or female, all of its mitochondria come from the mother.

This is a terrific boon for researchers trying to disentangle the genetic history of an organism. Mitosis (normal cell division) is a messy process, with swatches of DNA frequently "crossing over" between pairs of chromosomes. Because one of the chromosomes comes from the male parent and one from the female, it doesn't take many generations of mitosis for the distinct genetic inheritance from one parent to get hopelessly confused with that from the other. Not so with mtDNA: If it's there, it came from the mother—simple as that.

Crude attempts to trace the route of human evolution by counting the mutations in mtDNA were made in the 1970s, but it was experiments at the University of California Berkeley in the early 1980s that produced the first striking results. The researchers, Rebecca Cann, Mark Stoneking, and Allan Wilson, collected mitochondrial DNA from nearly 150 people from all regions of the world. Then, using 134 already identified "polymorphisms" along the mtDNA strand, they ran a computer program that compared each subject's DNA to every other's, grouping those with the fewest differences next to each other, those more diverse progressively farther away.

The result was spectacular. As one would expect, samples taken from Europeans tended to cluster near those from other Europeans, and so on. But deep down, near the roots of the "tree" showing the relative relatedness of all samples at the same time, the trunk split in two. Seven African samples occupied one branch; the remaining 141 sprang from the other. Assuming the rate of mtDNA mutation to be steady ever since chimps and humans diverged some 5 million years ago, that breakdown meant that all the people in the sample (Inuit, Rumanian, Australian aboriginal, everyone) shared a single common ancestor somewhere between 100,000 and 400,000 years ago and that this individual—a female, necessarily—had almost certainly lived in Africa. "Mitochondrial Eve" was born.

As with Mr. Targett and his cave-dwelling "ancestor," this isn't as impressive a result as it seems at first sight. The 148 people sampled have a common mitochondrial ancestor, not one common ancestor. Pick a different 150 people, and a different "Eve" would emerge. Anyway, considering the regrettably promiscuous breeding habits of our species, the pattern of links between every living human being and all others is so complex that it would take a computer the size of Rhode Island a few billion years just to sketch it.

Any number of scientists have attacked the methodology and data of the first "Eve" experiment, but repeated studies using improved data sets and software have produced results similar enough to convince most observers that the human species—*Homo sapiens*, that is—did originate in Africa not earlier than 500,000 years ago and (probably) not much later than 200,000 years ago and that, in a version anatomically indistinguishable from today's version, some portion of that population departed Africa somewhere around 100,000 years ago.

MtDNA is valuable for establishing differences as well as similarities. A collaboration between researchers at Penn State and the University of Munich in Germany caused a major stir in the human evolution community back in 1997 when they announced that they had been able to collect meaningful samples of mtDNA from a thighbone of a Neanderthal skeleton—in fact, the Neanderthal skeleton discovered in 1856.[1] Bad news for those who hoped that modern *Homo sapiens* and the prior inhabitants of Europe just merged quietly together genetically when they met some 50,000 years back. Not only did the Neanderthal exhibit more than three times as great an average difference between individuals as the modern human; it differed in quite different segments of the mtDNA genome. This suggests that the two strains began their drift apart half a million years ago, long before anyone would claim their common ancestor could have been remotely "human" in the modern sense.

Even though mtEve herself remains a cloudy and unsatisfactory figure, and the routes taken by her putative offspring on their conquest of the globe are too uncertain to rely on, she's no longer the only spokesperson available to give evidence. We inherit mitochondrial DNA only from our mothers. But there's another form of hereditary material with an exclusive sexual bias: the so-called Y chromosome. Only males carry a Y chromosome; indeed, the Y chromosome carries the particular gene that, matched with its partner on the woman's X chromosome, initiates the developmental processes that cause men to be male rather than female. The Y is a small chromosome—second smallest in the human set of twenty-six pairs—and many microbiologists suspect that a lot of even that little isn't good for much. The portion of the Y that is responsible for most of the maleness in males has a parallel swatch of DNA on the X to mate with, but a long tail of material at the opposite end of the Y appears to be pure evolutionary junk, coding for nothing useful and omittable without noticeable effect.

Yet another component is a set of highly variable repeated elements, short segments of genetic code in hundreds to thousands of consecutive multiple copies. This component also appears to have no effect on the viability of its host, but it does have qualities that make it valuable to genetic detectives. It may not matter when a single DNA "base" changes to another in one of these sequences, but once that has occurred, the change tends to stay that way.

One such change that has attracted particular attention is the one called "YAP," which resulted from a chromosomal rearrangement that almost certainly could have happened only once, with a 50% chance on the date being between 125,000 and 175,000 years ago. That's smack in the middle of the period where Eve's fans place her appearance, so the result attracted a good deal of attention when it was published in 1995 by Mike Hammer, an evolutionary biologist at the University of Arizona in Tucson.[2] Even better, according to Hammer, the YAP-type Y occurred most frequently in southern Africa, less often in northern Africa and the Middle East, and even less frequently in Europe. This suggests that however much his descendants have been back and forth since, "Y-chromosome Adam's" trip began in Africa.

Hammer got much more confused results when he tried to apply his technique to Asia—all to the good, as it's turned out. By tracking and comparing the subtly altered "tails" of junk DNA trailed by the YAP repeats, Hammer was able to distinguish not just one either/or choice at the given site, but five different "haplotypes," all occurring in difference percentages in populations from various parts of the world. Using this kind of data, Hammer and others were able to construct a crude map of the sequential peregrinations of humankind (the male variety, at least) of the last 50,000-plus years or so. The map shows first a wave of emigrants out of the west spreading across south and southeast Asia and into Australia, then (or simultaneously) a northerly wave into Europe. The next trace of our passage emerges far to the east, as the "Jomon" people settle into the Japanese archipelago from an unknown Asian direction. The Americas are settled next, according to these studies (more on that in a bit), with the well-established second, agricultural drift into Europe. Then, well after the beginnings of recorded history to the west, the islands of the Pacific are populated, and the Yayoi arrive in Japan from Korea to create the unique hybrid culture of Nippon.

*B*ut the most ambitious—some say grandiose—attempt to describe the origin, development, and geographic spread of modern humanity began before there was such a subject as molecular biology. Luigi Luca Cavalli-Sforza was a young M.D. from Italy's University of Milan when he fell under the spell of one of the greatest scientists of the century, Cambridge University geneticist Ronald Fisher who preached to a still dubious congregation of peers that the future progress of biology lay through the terrifying thicket of mathematics.

Fisher's first great contribution to the field was his theory of the origin, inheritance, and evolution of blood proteins: the already familiar "blood groups" A, B, AB, and O and the then-mysterious "Rh factor" first identified by a New York immunologist when Cavalli-Sforza was still a schoolboy. One of the intriguing things about the epidemiology of blood groups is that their frequency differs from one population to another, often along a clear geographic gradient.

Cavalli-Sforza was one of the first scientists to realize that tracing such gradients might also enable one to trace the evolution and ancient wanderings of our species. In the beginning, around 1960, all he had to work with were the crude blood group data already available and a primitive room-sized Olivetti computer. His first results, based on brute-force one-to-one comparisons between pairings-up of data from fifteen populations (three per continent), were published in 1962. They showed a tree with a first pair of branches diverging east and west, all Asians (and Native Americans) in one group, and Africans, Middle Easterners, and Europeans in the other.

With every advance in genetic data, statistical methodology, and computer power since that time, Cavalli-Sforza and his collaborators have refined and extended their model, and these refinements have produced no fundamental change in its general outline. They have become so supremely confident of its power to make fine discriminations between peoples and population movements that they trace everything in European prehistory from the Neolithic agricultural revolution beginning 10,000 years back to the comparative distribution of Indo-European and non-Indo-European languages (appearing 5000 years ago) and the expansion of Bronze Age "Greek" culture north and east from the Mediterranean (1000 B.C. or thereabouts).

Cavalli-Sforza has spent much of his career at Stanford University, which prides itself on encouraging interdisciplinary research. Like his colleague in the Stanford Linguistics Department, Joseph Greenberg, he is a generalist, drawing on data and theories from a wide range of specialized studies to provide ever more elaborate support for a theory that hasn't changed fundamentally for decades. A lot of younger scientists, who've come up through a far more specialized professional training, smile a little at the broad-brush assertions of such scholars, but few have dared attack the central citadel they've built. In science, it's a distinct advantage to be the one who, like Turner, Greenberg, and Cavalli-Sforza, "wrote the book."[3]

However much uncertainty and professional disagreement there may be over the results produced by biochemical–genetical–statistical methods, such techniques have enormously advanced the field of early-human studies. And the more capacious and refined their data sets become, the more unsatisfactory seem the kinds of analytic results that more traditional scientific anthropologists can hope to achieve with their materials. A striking example of the problem can be seen in recent studies of Paleo-Indian skeletal material by the distinguished Texas A&M anthropologist D. Gentry Steele and one of his former graduate students, Joseph F. Powell, now a professor at the University of New Mexico. The mathematical machinery deployed in the papers[4] is of impeccable lineage; it's their data that set alarm bells ringing.

The Steele–Powell studies are based on exactly the same kind of data that old-style physical anthropologists used (and sometimes misused) to assign the objects of their study to racial categories: measurements of the comparative length and breadth and facial characteristics of human skulls. And they draw on the same pitifully sparse data set that earlier researchers have had to deal with—even less data, in fact, because only eight specimens among the twenty-odd recognized sets of Paleo-Indian remains were deemed sufficiently complete to be useful for their approach.[5]

But instead of assigning single numbers to their specimens according to some simple formula such as "maximum width divided by maximum length equals cranial index," Steele and Powell poured the data from no fewer than eight separate measures into the mathematical mixmaster called statistical multivariate analysis, measuring similarity and difference not just between every pair of specimens, but between every pair of measured characteristics as well. Even with only eight data sets to draw on, the number of possible comparisons using this technique is greater than any scientist is likely to take on with pencil and paper or pocket calculator. Only the availability of powerful computers and even more powerful statistical software makes the calculations feasible and allows suggestive deductions to be made from the resulting flood of data.

The first conclusion produced by the Steele–Powell studies is not exactly earth-shaking news. "The general facial and cranial features of the oldest specimens recovered in North America affirm their anatomically modern appearance." However, the authors continue, skulls from their sample of 8000- to 10,000-year-old specimens "have a longer and narrower braincase than do most recent American Indians...."

So far, nothing that would have surprised Wormington or others of her generation. But with the calipers in their tool kit supplemented by computers, Steele and Powell had bigger, more elusive game in view. Drawing on measurements of thousands of old sets of remains from Japan, Peru, China, the Arctic, South Asia, and the Pacific islands, they performed another sequence of statistical prestidigitations ("principal component analysis of six common craniometric measurements corrected for size") to measure their Paleo-Indian sample's similarity to and difference from other populations.

Fortunately for nonstatisticians, they have also found a way to print out their results in a manner that makes them clear and concrete. Steele and Powell represent the whole known range of variation between human populations by a perspective drawing of a volume of space. The six different indices derived from their calculations are reduced by mathematically sifting to the three most significant, each index being read off on the chart as a distance, horizontal, vertical, and in (imagined) depth. The overall picture is one of a floating cloud of data points, denser toward the "center" of the space, more dispersed and irregular toward the extremes.

Because different symbols are used to represent groups of points— a cube for ancient material from the Chinese mainland, a flag for Okinawa, a pyramid designating the Paleo-Indian material, and so on— the range of variation between populations and the relative similarity between any two leaps out to the eye. Three things in particular are striking about the Steele–Powell chart. First is the great "distance" separating all the remains that represent ancient populations from one another: The Chinese skulls (from later areas of the same site near modern Beijing, where half-million-year-old "Peking Man" was found in the 1920s) and the Okinawa specimens are near opposite extremes of the diagram, with the Paleo-Indian sample well off to the side of both, so the shortest line connecting all three would follow a dog-leg path.

The second intriguing point is the spatial relationship between ancient groupings and modern populations. No matter what their geographic origins (Europe, Asia, the South Pacific, or the Americas), modern groups are grouped densely—that is, they resemble one another more closely than they do any of their ancient progenitors. They cluster in a complex "sheet" at the center of the cloud; the older samples lie isolated like wandering stars on the lonely fringes of the galaxy. The third striking fact that emerges from the study seems at first a veritable clincher to settle the heated debate over how, or even whether, the earliest inhabitants of North America are related to living Native Americans. "Surprisingly," note the authors, "recent North American Indians differ from Paleo-Indians more than any other geographic population, and [italics added] *it is the relatively short cranial length of more recent North American Indians that creates the difference.*"

Alas, for those in search of simple answers, the authors immediately point out a problem interpreting the data. Only two of the surveyed sets of remains, it turns out, were male, and for some unknown reason, skull proportions in the male specimens are nearer the contemporary

Native American average than those of the more numerous females. Also, looking beyond the tiny data set considered in their paper, they point out that in the few areas where there are enough old remains of various ages to compare (mainly in the Aleutians and in Tennessee), one can dimly see a drift over time from long-headed early inhabitants to more round-headed successor populations.

It's possible, of course, to claim that the reason for the drift is that an earlier, "Caucasoid," long-headed population was overwhelmed (or wiped out) by later waves of round-heads. Possible, but completely without evidential support. The problem with the Steele–Powell sample, as the authors tacitly admit in drawing their conclusions so cautiously, is that the sample isn't just small (so small that a single new good sample could radically change its implications). It is also drawn from isolated individuals who lived thousands of miles and thousands of years apart. As Tom Green, director of the Arkansas state archeology office, dryly remarks, "This in no way can be considered a breeding population."

Is Race Necessary? 8

"*A* breeding population...." Accustomed as we are to the idea that human beings are, biologically, animals, we still find it odd to think of ourselves in such clinical terms, as the outcome of a kind of genetic experiment in Mother Nature's laboratory. And there's no question that over the last few thousand years, the human animal has come to have the run of the lab, even deciding what direction the experiment will take. But if we want to understand what happened to our species on its long road out of Africa to world dominion, we must keep in mind the conditions under which it lived for all but the last fortieth of its existence.

Like all their close relatives—gorillas, chimps, and bonobos—humans are social animals: They cannot survive at all, let alone reach reproductive age, without years of support and instruction in the intimate society of others. The "society" in question is no political abstraction to be debated round the campfire, but a set of concrete, biologically determined facts. Alone in nature, a solitary human cannot reproduce and has little chance of continuing to live at all. A group of humans wholly dependent for their existence on foraging and hunting is limited in size by the carrying capacity of the land the group occupies.

All the evidence gleaned from hundreds of years of excavating traces of ancient humanity and putting together anthropological descriptions of surviving hunter–gatherers shows that for most of our early ancestors, most of the time, "society" meant a group varying from as few as ten to a high in the low hundreds, with the norm somewhere around forty or fifty. Hence the group tended to be larger than a single, linear family, but small enough that every individual could trace a blood relationship to every other. When we discuss what happened to our ancestors as they spread across the world, the unit of study is not the individual or the species, but the group: the interbreeding, interacting family-plus, impinging (sometimes peacefully, sometimes violently) on others like it in its endless search throughout the changing seasons for resources to sustain it.

Historically, much of anthropology and archeology consists of data, observations, and theory about the varied strategies adopted by such groups in their quest to survive and prosper. Only recently has science been in a position to speculate about the group as a biological entity, ever changing over time, and to investigate how conscious survival strategies interact with the mindless machinations of the "selfish genes" within.

The science equipped to pose and answer such questions is population genetics, and thanks to the work of mid-century researchers such as Haldane, Sewell Wright, and Cavalli-Sforza's teacher R. A. Fisher, it has become a powerful, subtle tool for illuminating issues in almost every department of biology. It is also, unfortunately, extremely difficult to master, requiring of its disciples the combined skills of logician and cryptographer, to say nothing of the lightning-calculator abilities of a card-shark. For those of us who found ourselves in deep enough water trying to follow Gregor Mendel's deductions about his damned sweet peas, population genetics is the open ocean in typhoon season. And to make things worse, at least one of the central ideas of the field as it applies to the subject at hand is maddeningly counterintuitive. This is the idea of "genetic drift."

*W*hen it became possible at last to study human inheritance in detail at the genetic level, investigators naturally concentrated their attention on cases where inheriting was bad news for the inheritors: the condition called sickle-cell anemia, for example, which occurs mainly in people of African descent, and the many variants of thalassemia, a similar disease that manifests itself in natives of the Mediterranean coastline. Such conditions were clearly inherited and, moreover, were inherited in population groups with a common geographic origin.

Because they were focusing on genetic anomalies that produce human suffering, geneticists didn't pay much attention at first to anomalous variations between individuals that seemed to have no effect on health or what is delicately called "reproductive success." Indeed, until there was a handy way of examining long stretches of DNA unit by unit, there wasn't really any way to *know* about many such variations. If a heritable difference didn't show up in an organism's physique or behavior—black versus blond hair, for example, or blue versus brown eyes—it was effectively invisible.

But gradually, it became clear that human genetic material could differ in manifold ways that were seemingly neutral and had no effect on the operation of the organism even at the most fundamental level. This was true not only of the aforementioned stretches of "junk DNA," but also in genes serving vital bodily mechanisms. The more deeply molecular biologists looked into the human genome, the more they realized that few if any genes come in just one flavor. Most occur in several more or less similar forms that can be distinguished via careful biochemical analysis, but which have little or no apparent effect on the lives or health of their possessors.

The ways in which such mutations come into being is itself a fascinating story, but their application, in studies of early humankind, involves what happens to them once they exist. It turns out that what happens depends to a large extent not on the mutations themselves (by definition, they have no effect, good *or* bad, on their carriers), but on the size of the population the individual carrier lives (and breeds) in.

At this point a little math becomes inevitable, but at least it's of a familiar type: simple random numbers, coin-tossing variety. Normal genes always come in pairs, except when they are separated briefly by cell division. But one kind of cell division is final: the kind where the result of the division is an egg or a sperm. Because one gene in each pair derives from the male parent and the other from the female, each egg or sperm gets one or the other, and *which* it gets is entirely a matter of chance.

Now let's imagine *one* particular couple and *one* particular gene on *one* particular chromosome in *one* of them; in this thought experiment, it doesn't matter which. For convenience, let's call the gene "Arthur." Now let's assume for simplicity's sake that our couple is and always will be strictly monogamous (which, by the way, population genetics itself has proved to be a very dubious assumption).[1] Very well. Male and female join in coitus, sperm meets egg, and a child is engendered. Quick: What are the odds that Arthur made the cut?

The answer is fifty–fifty, and it's fifty–fifty every time the same couple couples again. Thus, on the average, there's a 50% chance of Arthur turning up in one of their children, a 25% chance in two out of two children, one chance in eight in all three children of three, and so on, just like the odds of getting heads in so many tosses of a coin. And everybody knows that the more times you flip a coin, the closer the number of heads or tails approximates the theoretical ideal of 50%.

The problem in applying such statistics to human beings is that human beings don't go on flipping indefinitely. Hunter–gatherers tend to have a lot of children, but without the benefits of modern medicine, they also tend to lose a lot of children, so the two tendencies even out. Suppose three kids do reach breeding age. The odds are pretty good that Arthur will survive into one or more of *their* descendants: 8 to 1 in favor, in fact. But that one time out of the eight that Arthur *doesn't* make it, in all the future generations of the race, *there will be no more Arthurs.*

Suppose, more realistically, that Arthur isn't the exclusive genetic property of one individual. Suppose Arthur is distributed throughout a breeding population. Still, the same hard fact applies. There is *some* chance, however small, that Arthur won't be passed on. And the smaller the population, the more likely that is to happen. Paradoxically, exactly the same argument applies in reverse. The simple luck of the throw can lead to a generation in which Arthur becomes compulsory, turning up in every single individual. And again, the smaller the breeding population, the more easily it can happen.

A lot of things militate against incipient Arthurism or its reverse. Although hunter–gatherers do tend to live and breed in small, closely related groups, they also sometimes (through warfare, rape, or peaceful negotiation) bring fresh genetic material in to dilute the homogeneity. But the tendency persists. In theory, no matter how large and various the breeding population, the frequency of occurrence of any given gene is going to vary unpredictably, and sooner or later, in the long run, it will either become "fixed"—omnipresent and inescapable—or get lost.

This, in summary, is the idea of genetic drift: an inevitable random component blurring the tidy, sensible patterns of natural selection. But the general effect on a breeding population is clear enough: Over time, some genes (more often than not, those that don't occur with great frequency) are going to get lost in the chromosomal shuffle, while others (more often than not, the ones that are already very common) are going to become omnipresent. If one of these genes, or some coordinated combination of them, happens to code for a visible trait—a long, narrow nose, say, or bushy eyebrows—then pretty much everybody in the tribe (allowing for age, sex, and environmental factors) is going to have long noses and bushy eyebrows.

The simplicity of this pattern gets rapidly clouded when more than one tribe or group has to be taken into consideration. Except on the proverbial desert island, human beings are just not inclined to breed only within the tiny pool of people they grew up with. If there isn't another group to look for a mate in, people invent an artificial "other," creating the bewildering variety of clans, septs, moieties, and phratries that kept cultural anthropologists so busily employed for a century. Hunter–gatherer peoples, whose wide-ranging foraging habits frequently bring different bands into temporary contact, have both opportunity and motive to keep adding new blood and fresh genetic material to the common pool—and not because of any mystical awareness of the dangers of "inbreeding" (which is what the negative consequences of genetic drift and fixation were once called. Groups that swap breeding partners thereby become related in a way, and are hence more likely to help each other out in hard times—or, at least, are less likely to try to wipe each other out over a particularly choice rabbit warren or berry patch.

To cast light on such phenomena, population geneticists have resorted to mathematical models, with great success so far as members of their own tribe are concerned, but at the cost of leaving those not up to speed on the Hardy–Weinberg principle and the Gaussian box model somewhat in the dark. It was only recently that investigators began to realize that the subtle formulas devised by population geneticists to describe the evolution of micropopulations have consequences when applied to the most macro scale of all; the one-time, world-transforming process whereby human beings came to dominate the globe. To understand these consequences, we must embark on a more ambitious thought experiment than the one about a gene named Arthur.

Let's imagine, for simplicity's sake, a flat world without any such impediments to movement as mountain ranges and oceans. Also for simplicity's sake, it is a world without seasons, climate, or weather, stretching limitlessly out to the horizon. Now somewhere on its surface let's seed our planet with hunter–gatherer humankind (because the dish is infinite and the resources and hazards it contains are the same all over, it doesn't matter where). What happens over time?

At first, not a great deal. Our initial seed-stock lives and breeds and dies, wandering the surface of its world in search of the resources it needs to survive. Assuming that the birth rate exceeds the death rate by some fraction (it had better, or our experiment's in trouble), our initial tribe will in time exceed the number it can conveniently feed on the range it can cover. Not a problem. Resources are abundant out yonder; the tribe can subdivide, each moiety collecting resources on its own, and it can subdivide again as necessary.

So far, the tribes are behaving rather like cells in a petri dish, but in our world-sized petri dish the resources are self-renewing, so the spot where the initial infection was made continues to support life. Little by little, the infection spreads as the total population grows. A few groups may, through their random wanderings, find themselves far from the original breeding ground of the race, but also by pure chance, the total population will continue to cluster fairly near its point of origin.

Meanwhile, life continues on the genetic level as well. The sexes recombine and exchange their DNA, and that DNA continues to undergo its one-in-a-billion mutation rate. Little by little, the groups diverge, through mutation and subsequent genetic drift, the rate of divergence being retarded to the degree that groups exchange members and genetic information. Groups toward the center of the growing population have other groups on all sides as potential partners for exchange. But at any given time, some groups are going to occupy the margin of the territory inhabited so far. And as time goes on, the percentage of tribes foraging territories at any given distance from the margin steadily grows in relation to the total.

By the time our thought experiment has run for some thousands of generations, the original uniform genome of the species has accumulated a good deal of variation: There are Arthurs of every description scattered randomly across the map. Again, near the center of the distribution, any given Arthur has a fair chance of being transferred from one tribe to another. Out on the margins, though, where each tribe has far fewer neighbors, Arthur's chances of escaping the originating genome is smaller. The chance of an Arthur of the margins ever making it to the *opposite* margin, across the whole territory occupied by the race, is extremely slim, and it gets slimmer with every rise in population and consequent expansion of the territory.

Let's look only at the subset of Arthurs that *can* be looked at: those that have visible effects on their bearers. By definition, remember, Arthurs occur at random. In small populations, where gene flow between populations is low, some get eliminated by the luck of the draw, whereas other get "fixed," but *what* gets fixed is determined purely by chance: Bushy eyebrows are statistically no more likely than long noses, or green eyes, or prominent buttocks.

What *is* certain is that populations on the periphery, with less opportunity than average to dilute the genetic hand dealt them by chance, will tend to show a more distinct set of physical traits than tribes toward the center, and that groups at the extremes of the whole distribution will, on the average, differ a great deal more from each other than they do from the average.

If this argument is correct, then race—the great divider of humankind, the emotional and intellectual underpinning of endless theorizing and rationalizing at least since Herodotus came back to Greece from Egypt to found the science of descriptive anthropology—is an artifact of evolution, a meaningless side effect of the historical process.

Innumerable factors in the real world that human beings inhabit complicate the process described in the preceding paragraphs. In the real world, mountains and deserts and oceans radically change the odds of encounters between populations; new diseases emerge and old ones spread, wiping out whole peoples; environments change, technologies and ideas, which leave few traces in the archeological record, change too. These factors do not affect the fundamental process.

As it drifted out of Africa on its desultory path to conquest of the globe, humankind accumulated a great deal of genetic variation of little or no significance, beyond the cosmetic, to its possessors or to anyone else. Then, about 10,000 years ago, an invention made independently in several parts of the world turned the random genetic game upside down. With the development of agriculture came sedentarism and with *it* the concentration and emphasizing of "racial" characters through selective inbreeding.

Desert

Beulah Shows the Way 9

*I*t is a time-honored principle of Anglo-American jurisprudence that judges should not themselves make law. Under this system, it is up to plaintiffs to show how they have been injured and to identify the precedent or statute under which they seek relief. It is up to defendants to show that no injury took place or that no relief is to be had under law. The judge's job is to decide which side's arguments mesh best with existing law and practice, not to suggest other arguments that might be more persuasive or to decide the matter in accordance with some other statute entirely.

In the real world, or course, things are not so tidy. In the real world, and especially in equity (noncriminal) cases, judges enjoy a good deal of latitude in nudging a case in a direction that intrigues them. In his first decision in the case of *Bonnichsen et al.* v. *United States of America*, announced on June 4, 1997, and amplified in a 50-odd-page opinion issued June 27, Judge Jelderks cast a cold eye on the principal arguments presented him by both parties. But rather than tossing the case out of court entirely, the judge chose to suggest other grounds of argument that he was prepared to listen to with interest.

Jelderks's rejection of the scientist–plaintiffs' assertion that they had been deprived of their First Amendment and due-process rights under the Constitution verged on sarcasm.

[They] claimed to have been discriminated against on account of their race, i.e., because they were not Native American. Assuming, for the moment, that for purposes of NAGPRA Native American is a "race," as opposed to a political status, ... I fail to comprehend how plaintiffs are being treated differently ... There is nothing in the record to suggest that a Native American archeologist would be permitted access to the remains.... The tribes object to anyone studying these remains, regardless of race or tribal affiliation.

If the plaintiffs got the raised-eyebrow treatment, the judge was almost openly contemptuous of the Corps's claim that it had given the plaintiffs' wish to to examine the remains due consideration. He cited months' worth of voluminous internal correspondence showing the Corps as little concerned with anything but avoiding trouble with its Native American clients and as ignoring its own advisers' warnings that NAGPRA was a risky basis on which to justify its behavior.

Instead of dismissing the action, though, Jelderks handed down what might be called a Solomonic cop-out, satisfying neither party and leaving the case wider open than ever. The arguments so far presented on both sides did not hold water, he suggested, but his court would be the right place to continue the controversy—*if* the controversy were to focus on a number of points raised by the Native American Graves Protection and Repatriation Act, so far touched on only tangentially.

Judge Jelderks's list of points to be considered by the Corps numbered seventeen: a multipart essay question on a final composed by a professor from Hell.

In reaching its decision on the ultimate disposition of the remains in question ... the Corps should consider, inter alia, the following issues:

• whether these remains are subject to NAGPRA, and why (or why not);

• what is meant by terms such as "Native American" and "indigenous" ...;

• whether ... NAGPRA applies to remains or cultural objects from a population that failed to survive ...;

• whether NAGPRA requires (either expressly or implicitly) a *biological* connection between the remains and a contemporary Native American tribe;

- whether there has to be any *cultural* affiliation between the remains and a contemporary Native American tribe—and if yes, how that affiliation is established if no cultural objects are found with the remains;

- the level of certainty required to establish such biological or cultural affiliation, e.g., possible, probable, clear and convincing, etc....;

- whether there is evidence of a link, either biological or cultural, between the remains and ... any other ethnic or cultural group including (but not limited to) those of Europe, Asia, and the Pacific islands....

Behind the judge's carefully specific questions, a more general and certainly more inflammatory question obviously looms: Just what or who *is* a Native American? Jelderks's interest in arriving at an answer to that question emerges vividly in the footnotes appended to his list, which demonstrate extensive background reading and rumination on the subject. (They run nearly three and a half times as long as the text they annotate.) One note in particular reveals the judge's profound imaginative engagement in the case before him:

> A projectile point was found embedded in the remains, which may have led to the man's death. Defendants have suggested that the point was of a type formerly used by Native Americans, and cite this as proof that the man was an ancestor of today's Native Americans. They may be right. However, this also could be seen as proof that the man was *not* of Native American ancestry, but was part of a competing group—which might tend to explain how he ended up dead with a spear in his side....

"In other words, an early version of cowboys and Indians," the *Tri-City Herald's* Don McManaman deadpanned in his report at the time, "and the Indians won the first round." Both funny and just, McManaman's remark is also disturbing in that it points up the degree to which Judge Jelderks had fallen into thinking about the early history of North America along conventional "racial" lines.

By asking the Corps to clarify and define the terms of NAGPRA, the judge had implicitly accepted the specious but intriguing line of argument introduced by the plaintiffs' attorney Paula Barran in the very first hearing on the case back in October 1996: If no present-day Native American could show a plausible line of descent from Kennewick Man, then Kennewick Man himself could not be considered Native American. Given that on the testimony of James Chatters and others, the remains did not resemble those of modern Native Americans, the burden of proof was on the Corps to establish the connection. And the only way to do so would be through the very kinds of scientific testing demanded by the plaintiffs.

In principle, you could say it was a clean victory for the plaintiffs, though in practice, the judge gave the scientists and their attorneys little comfort. Their application to study the remains was denied. True, the Corps was sent off with its tail between its legs and a hefty homework assignment, but the remains would remain under lock and key, unexamined by anyone, for the time being. Meanwhile, the plaintiffs were instructed to collaborate with the Corps in seeking clarification of the judge's questions and in developing procedures whereby they could be answered. "The parties are to provide the court with quarterly status reports (preferably a joint report, but separately if they cannot agree)" at three-month intervals until "this matter is resolved."

Ten months had passed since the discovery of the remains, eight since the inception of the lawsuit. What had begun as a crusade looked like it was turning into a war of attrition, and some of the scientists who, in the heat and haste of the moment, had lent their names to the lawsuit were beginning to wish they had been more circumspect. It was embarrassing to be reminded periodically, by remarks of their own lead attorney to the press, just how much the case was costing. "The plaintiffs' legal bills for about 2000 hours of work exceed $250,000," wrote the *Tri-City Herald*'s Mike Lee in his first-anniversary wrap-up on the story, "though their lawyers aren't charging for their work. 'It would break them financially,' Schneider said, noting the scientists' universities aren't backing their employees' lawsuit." "This is really a David-and-Goliath case," attorney Schneider told Lee. "I mean we're going up against the federal government.... The only thing that is going to stop me is maybe a heart attack or exhaustion. We're in this thing for the long haul. There are just too many ... fundamental principles involved."[1]

When you're getting a free ride, it's impolite to pass judgment on the skills of the driver, let alone grab for the wheel. But it wasn't just occasional qualms about who might actually be paying for the gas that troubled the more unworldly plaintiffs: It was some of the fellow passengers they discovered on board with them. Having brought a suit arguing the right of scientific enquiry against claims of religious fundamentalism, they now found themselves allied in the same lawsuit with a religious body of considerably more bizarre provenance than any Native American tribe.

Only days after the anthropologists brought their suit against the Corps in October 1996, a Portland attorney named Michael Clinton approached Judge Jelderks and asked that the remains of Kennewick Man be "repatriated" not to the consortium of Native Americans contacted by the Corps but to his clients, a group calling itself the Asatru Folk Assembly (AFA). "If study shows this skeleton to be more closely related to Europeans than to Native Americans," stated the AFA press release at time of filing, "we think it should be turned over to us for proper study and for the proper religious rites."

At the time, some reporters assumed that the Asatru Folk Assembly must be an organization invented on the spot by libertarian mischief makers trying to discredit, by reductio ad absurdum, Native American claims to the remains. Far from it. The AFA has been around for more than two decades and today boasts "chapters" all over the country.

To a casual observer just another part-charming, part-pathetic offshoot of New Age "thinking," Asatru purports to be the revival of an ancient Northern European faith worshiping Odin, the Norse sky-god. Unlike similar pseudo-revivals, though, such as the cult of Brighid or the Wiccanist devotees of the Great Goddess, Asatru is very guy-oriented, playing up the warlike, weapon-wielding side of Druid–Viking fantasy. Attorney Clinton made his court appearances in traditional lawyer grab, but in a photograph much reproduced in the wake of the Asatru filing, he is seen in another guise: that of Nordicblond leader of the Portland-area Asatru offshoot, Wotan's Kindred. Here he sports knee-high thong-bound sheepskin leg-warmers, a sleeveless tunic bares his massive biceps, and he brandishes a yard-long broadsword.

Behind the religious rhetoric and the costume comedy, a reporter[2] for the civil rights group Coalition for Human Dignity discovered, both Clinton and Asatru founder Steve McNallen had associations fit to set any civil-libertarian's alarm bells ringing. Reporter Jonathan Mazzochi found that in a 1995 article in the AFA magazine *Wolf Age*, "McNallen gives a first-hand account of his time spent with the neo-Nazi group the Afrikaner Resistance Movement ... [He] describes his ... hosts as Africa's 'white tribe' ... [He] also writes that 'despite liberal misconceptions, much of South Africa had been occupied by white settlers before Bantu tribes migrated on the scene.'"

Clinton had some ideological baggage of his own to explain. Some two years before filing suit in Portland on Asatru's behalf, he had joined with Church of the Creator founder Ron McVan (motto: RAHOWA, which stands for "racial holy war") in hosting a Portland colloquium with David Irving, a scholar much esteemed on the extreme right for his denial that the Holocaust occurred. But Clinton was never called upon to explain his affiliation with Irving. The media treated Asatru's participation in the suit as a good joke, and the members' willingness to appear before the cameras in full Odinist regalia ensured them plenty of time on camera at their every appearance—to the intense embarrassment of the plaintiff–scientists, who found themselves unmercifully ribbed about their newfound comrades in arms by friends and enemies alike.

For the more conservative scholars among the plaintiffs, though, perhaps the most disturbing thing about the Kennewick case was its apparently inexorable transformation from a matter of hard science into a kind of serial drama with a life of its own. They were also uncomfortably aware that they themselves had, if not actively fostered that transformation, at least failed to speak out against it. In early statements to the press, as well as in the affidavits submitted to the court appealing for the opportunity to study the Kennewick remains, they had taken great pains to emphasize the scientific importance of Kennewick Man, in part because of the supposedly distinctive racial characteristics he exhibited. Having done so, they found it awkward to explain to wave after wave of reporters calling for newsy quotes that the story they were chasing wasn't quite as sensational or unambiguous as it seemed.

Now, it happens that as both plaintiffs and defendants parsed their way through Judge Jelderks's opinion looking for legal handholds, a scientific paper was grinding its way through the peer review process toward publication. The description and analysis of a discovery made eight years before in an Idaho gravel quarry, this paper might have served as a model of both solid, cutting-edge scientific archeology and sensitive, deft handling of political–cultural issues. Had the saga of Beulah been published in anything like timely fashion, all the fuss over Kennewick Man might never have occurred, or at least might have unfolded within more rational bounds. But the very obstacles that delayed its publication shed additional light on the peculiar status of archeology in North America today.

*A*s Jim Woods left home for his office on the campus of the University of Southern Idaho the morning of January 18, 1989, the forecast was for sunny, cold, and dry: pretty much an ordinary winter day for south-central Idaho and the valley of the Snake.

The day turned extraordinary shortly after 10 A.M., when the receptionist at the Herrett Center for Arts and Sciences notified Woods that there was a lady on the line calling from the nearby agricultural community of Buhl. The caller was a Mrs. Burkhart, a teacher at Buhl High who said that she had visited the Center's exhibits on High Plains prehistory with her students and thought the staff should know about some bones her husband Nellis had spotted the day before while working the rock crusher in a county gravel pit. He'd brought the bones home with him that night, and she had taken them to school the next morning to show the biology teacher. They were human, he said, no question about it. Did the Herrett people want to take a look? No, she didn't think her husband would mind if they went out to the pit and talked with him.

Asking his collections manager Phyllis Oppenheim to grab her notebook, a camera, and some boxes and wrapping papers in case there was anything to bring back, Woods drove the sixteen miles west from Twin Falls to Buhl, turned right at the Green Giant packing plant, and followed his informant's directions to the gravel quarry on a shallow slope overlooking the Snake a quarter-mile or so to the north. Parking just off the road, the visitors bumbled their way through the growl and dust of a dozen maneuvering trucks and backhoes until they spotted the sorter where the remains had been found.

Its operator was not exactly glad to see them. Idaho is a state of outdoorsfolk and curio collectors, but it also is a state with very strict rules governing the discovery and proper treatment of archeological materials, and rural legends abound about the grandmother of the friend of a friend busted for picking up an arrowhead by the roadside. Burkhart's main fear, though, seemed to be that the visiting scientists might want to call a halt to the work.

Assured by Wood that work could continue, Burkhart conducted the scientists to the spot where the loader had been operating the day before and left them to see what they could find. There might well not be anything left, he warned; he hadn't been looking out for bone on the conveyor belt, only for material that could jam the crusher, and a lot of other stuff might have got by among the mixed boulders and gravel before he noticed.

By 11:30 Oppenheim and Woods were about ready to agree. Conditions were terrible. The ground beneath their feet was churned up by the treads of earthmovers; the gravel pit wall loomed over sixty sheer feet above their heads, layer after layer of gravels of every shape and size, larded with lenses of sand and, a dozen feet up the overhanging wall, gigantic boulders deposited in the megaflood following the Ice Age collapse of glacial Lake Bonneville.

It was Oppenheim, stumbling awkwardly along the rubble slope at the foot of the bank, who spotted only inches from her nose a minute sliver of white, no greater in diameter than a pencil eraser. A little delicate brushing and troweling revealed a tooth, unmistakably human, still attached to a lower jaw embedded upside down at the very base of an accumulation of river gravels more than sixteen feet deep.

Turning from the bank, the inspectors discovered that they themselves were under inspection: Workmen were taking advantage of their lunch break to check out what the aliens were up to. Woods and Oppenheim agreed they needed both equipment and reinforcements. Leaving his assistant to stand watch, Woods rushed back to Twin Falls, picked up an excavator's field box, and put in an urgent call to the Historical Museum in the state capital, Boise, and his friend the state archeologist, Tom Green. By the time he got back to the pit, Oppenheim's audience had increased to a dozen or so; word was spreading that something significant had turned up, and traffic on the busy county road was slowing to a crawl as it passed the quarry.

By the time Green covered the two-hour drive from Boise, Woods and Oppenheim had managed to expose eight or ten skeletal elements roughly aligned in what seemed to be sandy gravel. But it was already clear that quarry work and curiosity seekers weren't the only threat. As the day faded toward afternoon, the low sun had come around to shine directly on the gravel face, melting the frost firming the earthy overhang atop the sixty-foot bank.

Taking charge of the enhanced team, Green decided that even if the insecure bank didn't collapse during the long refreeze of the winter night, there was no way to maintain security on the site. (By late afternoon, the crowd watching topped two dozen, including commuters on their way home to the nearby Kanaka Rapids subdivision and some high schoolers with a video camera.) The work had to be finished before nightfall.

But by sunset at 5:30 P.M., the diggers still hadn't found a skull. The team pulled their cars up to the base of the cliff on the now quiet quarry floor and trained their headlights as best they could on the worksite. Then, shortly before 7:00, with the near full moon climbing the eastern sky and amid the occasional flash of an onlooker's Instamatic, the cranium emerged from the gravel, over four feet from the spot where the jawbone had been found. At 7:00, the excavators ceased working, their total bag the cranium, the mandible, some ribs, some vertebrae, upper limb bones, and a femur, as well as a number of artifacts suggesting that the remains were those of a formal burial. It was cold comfort that they had almost certainly found all that remained to be found: Considering that the femur was actually discovered on the conveyor belt of the crusher, the lower part of the skeleton had very likely been reduced to powdered chalk the day before.

It was immediately clear to Tom Green that the remains found at Buhl Quarry were old—probably, given their position deep beneath many yards of river deposits, extremely old. Nevertheless, his first call, upon returning to his Boise office Wednesday morning, was not to the governor's press secretary or the news desk of the Idaho *Statesman* but to the Cultural Resources office of the Shoshone-Bannock Tribes at the Fort Hall Reservation near Pocatello, some 250 miles to the east, near the Wyoming border.

As state archeologist, Green was specifically charged under Idaho law to consult with the state's Native Americans about the handling of ancient remains. The first major find to turn up in Idaho since the law went into effect in 1984, the Buhl remains would raise a lot of expectations and set a lot of precedents, and Green wanted to be sure no misunderstandings occurred. The tribe affected was one of the more militant about any disrespect shown the remains of its "ancestors."

Green told his Native American contacts that at a first guess, the Buhl remains appeared to be at least 5000 and perhaps as much as 8000 years old and that artifacts found with the skeleton—a two-edged obsidian tool, parts of a bone needle and punch, and the penis-bone of a badger—suggested a formal burial. After some discussion, the tribal elders notified Green that he was authorized to hold on to the remains for further measurement and study and—most important—gave him permission to send a portion of bone out for radiocarbon age dating.

Permission to study was all very well, but who was going to pay for it? Not the tribe, and certainly not Twin Falls County, owner of the gravel pit. Green got in touch with R. E. Taylor's Riverside radiocarbon lab, which agreed to perform the dating free on the understanding that the lab's own work and that of other campus departments would take precedence. Green agreed to wait. But as the months passed without news, he and his Indian contacts began to grow impatient. Then an earthquake damaged the Riverside equipment, and more months ticked away while the lab worked to recalibrate the delicate gear. Finally, Green felt he could wait no longer. The specimens were returned from California and sent off to a lab in Switzerland that promised fast action. Going on three years after the discovery of the remains, Green was thunderstruck when Eidgenossische Technische Hochschule of Zürich returned its results: 10,675 years before the present, with a very narrow uncertainty of less than 100 years either way.

Meanwhile other studies were proceeding, as personnel and time were found to devote to them. A geologist at Walla Walla Community College managed to spend some time at the quarry in early 1992. By that time another ten to fifteen feet of the gravel bank where the bones had been found was gouged away, but Bruce Cochran was still able to decipher the lay of the land when the burial took place. The body and its gravegoods were laid to rest in a wedge of fine sand deposited in quiet water along a gravel bar. Not long afterward (the sand was hardly disturbed by burrowing animals), a rush of high water spread gravel above the grave, where windblown sand and loam quickly filtered into the crevices.

Though all the bones below the lower ribs were lost, there were enough remaining from the upper body to reveal that the remains were those of a woman in young adulthood—somewhere between seventeen and twenty-one at the time of her death—and to provide an astonishing amount of information about her short life. Both bones and teeth showed that although she was healthy when she died, she had experienced numerous and prolonged periods of undernourishment. Analysis of the the bone protein collagen showed that she ate both meat and fish, and tooth wear indicated that her diet probably included processed, cooked vegetable matter as well; the only likely explanation for such severe abrasion was grit from a volcanic-stone *metate*, or grinder.

By the time Green and his seven co-investigators managed to get their report into print nearly nine years after the discovery,[3] the brouhaha over Kennewick Man was already under way, so the section of the report dealing with ethnic affiliations (inevitably, once it was known that the remains were female, they were dubbed "Buhl-a," which, just as inevitably, soon became Beulah) was extremely cautious and understated. Todd Fenton, a grad student at the University of Arizona, was willing only to state that Buhla's skull "exhibited a suite of craniofacial features that fall within the range of American Indian or East Asian populations." Because of the extreme erosion of the tooth surfaces, most of the traits considered ethnic telltales in Turner's catalog were not to be seen, but what characters were evident confirmed the general East Asian–American Indian diagnosis.

More study might have yielded more definite information, but as it happened, Fenton had been lucky to make any measurements at all. When Green received the ancient date from the Zürich lab, he approached the Shoshone-Bannock for permission to conduct further studies, but this time the elders weren't having any. Study had already dragged on more than two years longer than had originally been expected, they told him. Recent deaths on the Fort Hall reservation had stimulated speculation among tribal members that the Buhl Woman's spirit was responsible, demanding the peace that only reburial of her bones could give her. Fenton's examination took place in three marathon days of measurement, photography, and X-raying, while Green and the elders worked out a repatriation plan. In mid-December 1992, Green transported Buhla and her gravegoods to Fort Hall for reburial and sadly drove back home.

For his efforts to extract the greatest possible amount of information from the Buhl burial while radically constrained by issues of money, time, and politics, Green was treated by many of his archeologist colleagues as a traitor to science. The most extreme assaults came from physical anthropologists led by Clement Meighan, a retired professor from UCLA, whose American Committee for the Preservation of Archeological Collections successfully brought suit in California to prevent repatriation of more than 10,000 skeletons and funerary objects in state collections. "They're throwing away one of the two or three major finds in the New World," Meighen told a reporter for the environmentalist newsletter *High Country News.* "We're talking about a skeleton that was around 5000 years before the pyramids of Egypt were built. Repatriation is a loaded and improper term because it implies that you're giving something back to people who own it. They don't own it, and never did."[4]

Green, now director of the Arkansas Archeological Survey, has no regrets about his handling of the Buhl case. "We've seen what happens when archeologists' first thought on making a discovery is to call the *New York Times.* It seems to me more and more that to do archeology at all you have to be a self-promoter, and the Paleoindian field is great for that kind of thing, the earliest this and the most unique that. America's not the only place you get that; look at the battles between Richard Leakey and Donald Johansen over who gets access to the oldest bones. But you can't do science in the middle of a circus. From the beginning I said to myself, 'I'm not bringing that kind of thing to Idaho.'"

For all the limitations under which it was created, the report on Buhl Woman that emerged more than five years after her remains were returned to the earth is perhaps the most vivid, detailed, and colorful portrait of an ancient inhabitant of the Americas yet produced: evidence, if any were needed, that serious science can be done while respecting both the law and the beliefs of Native Americans. But as the Kennewick case demonstrated, scientific evidence can be ignored when it does not support the political agenda of determined partisans. Here, the kind of media bonfire that Green had worked to damp down in Idaho was deliberately whipped out of control, and this time no central responsible figure tried to put it out.

After nearly a year with no resolution or even any apparent progress, the Kennewick case only got more tabloid-sensational. Even at the putative top of the media food chain, it began to acquire a conspiratorial tone: "Is it possible that the first Americans weren't who we think they were?" wrote anthropologist (and thriller author) Douglas Preston in *The New Yorker.* "And why is the government withholding Kennewick Man, who might turn out to be the most significant archeological find of the decade?"[5]

In fact, a good many established figures in the field deplored both the exaggerated claims and the extravagant tactics employed to make the scientific case against repatriation. But the ferocious *public* controversies that marked turn-of-the-century early-American studies have long vanished. Willing as today's anthropologists and archeologists can be to go for the throat in private disagreements, they are rarely pass public judgment on one another, even in high-profile cases.

About the time Preston's *New Yorker* piece appeared to tell once more the Standard Version of the Kennewick Man tale, the first hints of a coherent counterattack were beginning to take shape—not through the public efforts of reputable scientists but at the behind-the-scenes behest of a politician named William Jefferson Clinton. Thanks to the ponderous machinery of government, however, it would still take the better part of a year for the President's initiative to show perceptible results. In the meantime, the comedy of errors continued.

Odin the Thunderer *10*
Meets
Jean-Luc Picard

*A*s a policy, the Native American Graves Protection and Repatriation Act was something quite new. As a piece of legislation, however, it was only too typical of the slapdash American Federal style: a mishmash of grand generalizations and niggling specifics, with virtually no attention paid to how the two extremes were to be reconciled in practice.

Typical also was the fragmentation in responsibility for implementation of the law. Each federal department affected was left to develop its own guidelines and practices under the ultimate responsibility of its director—the Commander of the U.S. Army Corps of Engineers (for land under its jurisdiction), the Secretary of Energy (for sites such as Hanford), the Secretary of the Interior (for the Bureau of Land Management, Indian reservations, national parks, and so on)—with no provision made for joint policy development or any thought given to the advisability of same.

The result was another round in the game of administrative pass-the-buck that makes assigning responsibility for anything whatsoever under federal jurisdiction such an infuriating exercise. Rather than devoting any resources or attention to developing a policy, most of the affected agency heads did what the Corps commander did: delegate enforcement responsibility for NAGPRA down the organizational chart. And it was delegated and redelegated until it reached someone who couldn't dodge the bullet, the officer "on the ground." In the K-Man case, this was the Walla Walla district's Colonel Donald Curtis.

Colonel Curtis stepped manfully up to the spitball that fate had pitched him. But neither he nor the representatives the Department of Justice mobilized to make the government's case realized until far too late exactly what kind of game they were playing in and the kind of tactics the opposition was prepared to use to win. For a solid year, a couple of small-town lawyers for a few "unworldly" academics walked all over the federal team, to the boos of commentators and crowd alike.

The Corps' public relations team celebrated the first anniversary of K-Man's discovery with their biggest gaffe yet. On July 30, 1997, Walla Walla spokesperson Duane "Dutch" Meier admitted that, far from sequestering the Kennewick remains from all comers pending Judge Jelderks's decision, the Corps had unlocked the storeroom containing them on no fewer than five separate occasions, each time to permit one or another Native American group to perform what Meier referred to as a "spiritual observance." Attorney for the scientist–plaintiffs Alan Schneider professed himself shocked by the Corps' failure to adhere to its own rules. "It's almost enough to make a lawyer speechless," he told the *Tri-City Herald*'s Mike Lee.[1]

The trouble was only beginning. Having revealed that it had acceded to Native American appeals to conduct religious services over the remains, the Corps was forced to negotiate with the Odin-worshiping Asatru Folk Assemby for equivalent visitation rights, and a private ceremony was duly held in the conference room next door to the Pacific Northwest National Laboratories storage locker. A separate public rite was also celebrated on the riverbank discovery site for the benefit of the public and the media. The comedy of errors continued through the next few weeks, when comments about the condition of the remains by Asatru lawyer Michael Clinton made it clear that, through ineptitude or design on the part of the Corps, the bones the cult had prayed over weren't K-Man's at all but others that had been found a month later along the Columbia bank—ones that were unquestionably Native American.

Meanwhile, other voices were being heard among the media cacophony. The same day Asatru leader Steve McNallen was intoning to the thud of his drum at PNNL, Washington State University geology professor Gary Huckleberry announced that he intended to ask the Corps for permission to dig a hundred-foot trench perpendicular to the Columbia in order to get a clearer idea about the geologic context of the site. If any artifacts or remains were encountered, Huckleberry added, they would of course be left strictly alone.

Because the odds of such permission being granted without Native American protest—particularly in view of the fact that Huckleberry proposed one James C. Chatters as collaborator on the dig—approximated those of a snowball surviving a summer in hell, spokespersons for the Corps did not encourage anyone to entertain false hopes. But the mere proposal, coming from a reputable scientist unaffiliated with the plaintiffs, was just one more negative blip for the Corps to cope with.

If the Corps still believed that hunker-down-and-take-it would ultimately carry the day, patience was rapidly running out in other quarters. Shortly after Labor Day, the office of Secretary of the Interior Bruce Babbitt received a call from a high-ranking member of the staff of the White House Office of Science and Technology Policy. The boss, an avid newspaper reader, was getting increasingly concerned about this Kennewick business, the caller said. The Corps of Engineers appeared wholly unable to make the federal case, the administration was beginning to look unreasonably partial to Native Americans and hostile to science, and there were individuals on the Hill only too ready to make political hay of that. Wasn't Interior supposed to be advising the Corps on this issue, for goodness's sake? What was being done? Well, do it.

As a matter of fact, the Corps' archeologists, faced with Judge Jelderks's seventeen trenchant questions, *had* found themselves out of their depth and, at the urging of their frustrated handlers at the Department of Justice, had gotten in touch with Interior, where work had been quietly going on for almost five years to develop standard practices for handling inadvertent discoveries like Kennewick Man.

Interior found itself in an awkward spot. It could provide the Corps with desperately needed expertise and counsel—indeed, it already had agreed to do so informally by helping the Corps develop answers to Jelderks's seventeen queries, soon due on the judge's desk in the first of the quarterly reports mandated by his June 27 opinion. But Interior had no way of ensuring that the Corps would take its advice or use that advice wisely. Was it really a good idea to take a chance on getting tarred with the same brush being applied so dexterously to the Army? What decided the question was less policy than politics: In the executive branch, one just doesn't say no to the President.

A coordinated approach to the Kennewick defense slowly began to come together as summer turned to fall, with experts from the National Park Service (the subunit of Interior most experienced in NAGPRA affairs), the Corps, and Justice drafting, circulating, and redrafting a joint position paper that might satisfy the judge. Their task was not simplified when, in mid-November, the Honorable "Doc" Hastings, Republican of Pasco, introduced a bill in the House of Representatives essentially rewriting NAGPRA, ensuring scientists free access to study "unaffiliated" remains discovered since the law's passage, and essentially rendering the case in Jelderks's court moot.

Meanwhile, the Corps was grappling with the request of WSU's Huckleberry to trench the K-Man discovery site. Not trusting this time to a flat-out refusal, the Corps spent four months patching together a plan whereby Huckleberry would be permitted onto the site, but only to examine and take soil samples from the riverbank, and then only in collaboration with other excavators from the Corps and from the tribes.

Huckleberry had filed his request just before Labor Day. He was finally allowed onto the foggy, freezing riverbank the week before Christmas. As much time was spent clearing away knotty, deep-rooted undergrowth and answering questions from dozens of journalists from around the Northwest as taking soil samples from the bank and sifting cold river muck.

Once more it was Jim Chatters, invited (over vehement Indian objections) to participate as part of Huckleberry's team, who was able to supply reporters with the only real nugget of news. Arriving at the site early one dark morning before the rest of the crew, he managed to spot another fragment of human bone in the neighborhood of the original find two summers before. It had been churned up from the sediments, he suggested, by waves from the passing powerboats in the annual Tri-Cities Christmas cruise.

The Corps didn't earn any points with press and public with its belated, grudging permission for riverbank study, and with its next set of moves, it very nearly threw away the game. Immediately after New Year's Day 1998, attorney for the plaintiffs Alan Schneider revealed that he had learned the Corps was proposing to helicopter tons of rock to the Kennewick Man discovery site, drop it along the riverbank, and landscape the area with new, dense vegetation.

A Corps spokesman conceded that such a plan existed and that implementation might well begin immediately, but he said its only intention was to "secure" the site from further erosion and possible souvenir hunting, not to destroy it. He might as well have held his peace. Before another week had passed, Washington state's Senator Gorton and the Tri-Cities' Representative Hastings were writing jointly to the Corps to protest the plan to "bury" the site. "We can see no reason to intentionally damage the site by depositing tons of rock and soil," they maintained in a letter to Colonel Curtis (and faxed to every newsroom on their respective media lists). "Surely there are other precautions the Corps can take to protect against erosion that cost far less and don't have the same irreversible negative consequences."

What irreversible negative consequences? The legislators' letter left them undefined, but another letter, produced by Schneider as though on cue, also affirmed that "critical data" about the site would be lost if the plan went forward. "Stabilizing the area with feet of boulders and dirt is destruction of evidence," wrote University of Colorado geologist Tom Stafford, the scientist tapped by the plaintiffs to perform further radiocarbon dating analysis on the Kennewick remains should they ever be given permission to conduct such an examination.

As the sun returned to the frozen Columbia Valley, and while the Corps' press-savvy antagonists made mincemeat of the engineers' every attempt to explain and justify their actions, there were plenty of other topics to keep the Kennewick Man story alive. In early February the ever-resourceful Chatters unveiled a portrait bust of Kennewick Man by Richland artist Tom McClelland, sculpted to Chatters's specifications and immediately recognizable as a near relative of the good Starship *Enterprise*'s Captain Jean-Luc Picard. First published in a special *New Yorker* spread, the photo was immediately reproduced in media worldwide, renewing the widespread fascination with the K-Man story by giving the anonymous bones a "look" and a personality, however speculative. Once again the Tri-Cities Chamber of Commerce began receiving calls from news organizations and film producers asking for information and assistance on upcoming shoots. It was rumored that CBS's venerable "60 Minutes" was preparing a segment. Could a mini-series be far behind?

Not far at all, as it turned out. About the time that the *60 Minutes* crew wound up its shooting, the Tri-City Chamber of Commerce's Karen Miller got a letter from Anna-Claire Schröder, doing some advance work for the London-based independent production company RDF Television. RDF had been commissioned to produce a full-dress documentary for Great Britain's "alternative" national commercial network, Channel 4, Schröder wrote, and would be arriving in Kennewick in early July to film interviews with the principal players in the case and to shoot background shots of town, river, and desert.

Unfortunately, RDF's schedule wouldn't permit the crew to stay around till the end of the month to shoot the crowds and excitement of the Water Follies, so would the Chamber be kind enough to arrange for a few of those cute little hydroplanes to be present on the river during the Fourth of July weekend to provide some exciting backgrounds for RDF's shooting? "The film would just not be the same without the boat race, which plays a pivotal part in the narrative," Schröder wrote.[2]

Miller replied to RDF tactfully as she could, explaining that unlike other down-home American sports like dirt-bike racing or tractor-pulls, unlimited hydroplane racing was a serious, expensive, and rigidly scheduled activity, and that those cute little boats (weighing a ton or so each) would be busy with the Madison Regatta on the Ohio River over

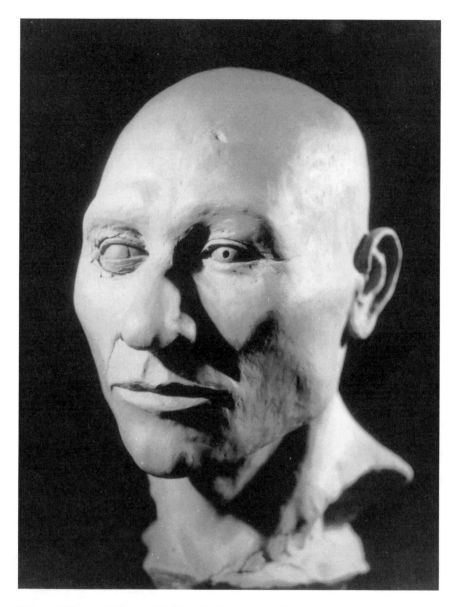

"Portrait" bust of Kennewick Man by Tom McClelland. (Photo by Max Aguilera-Hellweg.)

the Fourth. Her tact did not prevent her from mentioning RDF's request to *Tri-City Herald* reporter Mike Lee, who in a mid-June column had a modest amount of fun at RDF's perception of hydro reality.

The hydro request wasn't the only indication that RDF had some curious ideas about how to operate in the wild and woolly American west. RDF producer Eve Kay dropped a line to Judge Jelderks in Portland announcing her impending arrival, and asked when it would be convenient for the judge to make an appearance before her crew's cameras. The judge did not reply enquiring, as he well might have, if British magistrates routinely gave interviews to the media concerning cases currently in litigation. Indeed, he did not reply at all. But Kay's letter found its way onto a number of bulletin boards in Portland's Federal Courthouse, where it afforded the staff a chuckle or two. "It looks like she got her idea of American justice from *The Life and Times of Judge Roy Bean*," one docket clerk said.

By the time RDF's documentary made it onto U.S. TV, courtesy of cable's Discovery Channel—under the title "Homicide in Kennewick"—anticipation was running high. But despite some dubious "re-enactments" in the manner of tabloid TV, the show proved almost disappointingly sober and even-handed in its presentation of the legal and cultural issues involved in the Kennewick case. It was certainly Jim Chatters's best platform yet: as it turned out, the best he was ever to possess. By the time the show appeared, the story, both in court and in the media, had moved onto less dramatic and more problematic ground.

Why had it taken the authorities a year and a half to discover that Chatters's inventory didn't match Longenecker's? Largely because it didn't occur to anyone to make a comparison. But once a comparison *was* made, there was no question that the difference was real and highly significant. The bones not mentioned in Longenecker's list were not just a few small fragments of rib or digit but substantial chunks of both of the skeleton's femora: bones on the scale of that in a bone-in ham and not to be confused with any other in the human body, even by an amateur. Although Chatters's written inventory had not specified how many fragments each femur had been found in, the videotape and slides made shortly before he turned the remains over to the sheriff showed the bones clearly. Unmistakably two, and perhaps as many as four, pieces simply did not appear in the list made by Longenecker only weeks later.

Before the full dimension of this discovery could be absorbed, the press spotlight was wrenched back to the question of whether the discovery site would or would not be buried. On March 12, the Corps announced it was ready to proceed. Three days later, the plan was on hold, pending approval from authorities of the National Register of Historic Places. The very next day, at the behest of Senator Gorton, a Senate committee marking up an emergency disaster-relief bill inserted language forbidding the Corps to spend any money stabilizing the site. Before the week was out, the National Register, famous for taking months to answer letters, had given the Corps permission to carry out the work the Senate had just forbidden, a contract for the work had been issued, and the first truckloads of dirt and rock had begun to accumulate in Columbia Park.

Ten days later, opponents of the site stabilization plan were jubilant when the Corps abruptly announced that in response to opposition in both houses of Congress, the stabilization plans had been put on hold until further notice. They would have been wiser to have noted the date on the Corps' press release: April 1. At 7 A.M. the following Monday, the first tons of rip-rap were being lowered into place along the riverbank. Once again, the Corps had demonstrated its command of the situation, at the expense of appearing deceitful, obstructive, and wholly indifferent to the interests and concerns of anyone beyond its own chain of command.

As a matter of fact, even before the rocks began to splodge down into the Columbia mud, the Corps' iron grip on the Kennewick case was slowly being pried loose. Back in March, under pressure from the White House, the top command of the Corps had already reluctantly agreed to transfer principal responsibility for dealing with the Kennewick remains to the Department of the Interior—just in time for the rockets set off by the Corps' latest public relations gaffes with Congress to explode in Secretary Babbitt's office rather than the Pentagon.

On May 8, 1998, it became official: The name of Secretary of the Interior Bruce Babbitt was added as a defendant in the action pending in Portland Federal Court, and National Park Service chief archeologist Francis P. McManamon was put in charge of cleaning up the mess.

*A*rcheologists are not, on the whole, highly sociable human beings. Archeology's grand old men (and until Marie Wormington came along, they all *were* men) tended to be on the crusty, intemperate side, far more comfortable in the field issuing orders to toiling "natives" or grad students than around the committee table trying to appear interested in the views of others. Many of the older school have found it difficult to adapt to the contemporary realities of the profession, in which governmental bodies make the rules for their once free-wheeling profession, and ends are to be attained only through endless consultation.

At first sight, Frank McManamon would be no one's idea of ideal casting for Indiana Jones. His voice, posture, dress, and demeanor suggest rather the role of assistant professor or law librarian; they certainly offer no hint of a high school jock who went to college on a football scholarship or a veteran of twenty years climbing the bureaucratic ladder. Even for that lot in life, his manner seems remarkably mild. But so, one remembers upon making closer acquaintance with McManamon, was Clark Kent's.

Thus far, the fringe of the spotlight had barely brushed McManamon. Some unusually diligent members of the press had read the overview of NAGPRA and government archeology policy submitted by the government in January (three months late) in answer to Judge Jelderks's demand for a Corps response on some of the issues raised by the plaintiffs. McManamon's name was on that report, but because it was submitted in support of the Corps' position, it was easy to miss the fact that it represented a radical shift in the government's position on the case.

The document offered some comfort to both parties to the suit. On the one hand, McManamon opined, there was nothing whatever in NAGPRA forbidding scientific examination of remains in order to establish "affiliation" toward possible repatriation. Score one for the plaintiffs. But, said McManamon, NAGPRA clearly applied to *all* human remains discovered on federal lands, not just to those recent enough to offer reliable evidence of lineal descent. A family tree was not the only grounds for establishing affiliation between remains of the deepest past and living inhabitants of the land. As for a need for Congress to amend or elaborate on the 1990 statute, McManamon saw no necessity. The statute was clear enough as it stood; all that was necessary was to interpret it fairly, firmly, and fearlessly. He immediately set about doing just that.

The plan McManamon devised to reconcile the legal and scientific issues in the Kennewick case was to go through numerous revisions before being issued in final form,[3] but the strategic outline is visible even in early drafts. To begin with, McManamon implicitly granted the plaintiffs' argument that the government lacked the information it needed to decide the proper disposition of the remains. "Based on a review of existing documentation of the remains, the disturbed context in which they were found, and the manner of their collection," he wrote, "additional examination, recording, analysis, description, and interpretation of the human remains and about the archeology and history of the Indian occupation of the Kennewick area are needed to resolve the issues."

But there was no comfort for the plaintiffs in this concession. The authority cited for McManamon's statement was the hated NAGPRA and its predecessor ARPA, whereas the plaintiffs were denied any interest in the matter—indeed, they were denied so much as a mention. "[The Department of the Interior] shall undertake the necessary historial and scientific investigations," McManamon continued, "in consultation with Indian tribes that may be affiliated, and in cooperation with the Corps of Engineers, Department of the Army, and Department of Justice."

Despite their being acknowledged as intimately concerned in the matter, McManamon's plan offered little to satisfy the Native Americans' position on Kennewick, either. "If the remains are found to be Native American, as defined by NAGPRA, subsequent investigation and related studies will be undertaken to provide background for determining the ultimate disposition of the remains." But, McManamon warned, "If the remains prove to be as ancient as suggested by the radiocarbon date that was obtained on one bone fragment, it is not possible for any relationship of lineal descent, as defined by NAGPRA, to be made."

Why? McManamon pointed to a fact much obscured by the Corps' early decision to repatriate. A number of culturally and linguistically distinct tribes seasonally ranged the banks of the Columbia River in early historic times, primarily to take advantage of its annual spring and summer salmon runs. Most of these tribes at one time or another signed agreements or treaties with the Indian Land Claims Commission that acknowledged "traditional" boundaries and other territorial claims, usually to the advantage of the white settlers. But, McManamon wrote, "A careful legal analysis of the judicial decision by the Indian Land Claims Commission and the Court of Claims shows that the land where the remains were discovered has not been judicially determined to be the exclusive aboriginal territory of any modern Indian tribe." Therefore, even if the remains were determined to be Native American under NAGPRA, NAGPRA's territorial provisions wouldn't suffice to settle the repartiation question. Once again, further study— of "a wide variety of geographical, kinship, biological, archeological, linguistic, folklore, oral tradition, historical, and other information"— would very likely be needed to do that, and even so, it might well not be sufficient.

Nor were Native American sensibilities to receive more than a due measure of consideration. Further examination of the remains would be necessary even to fulfill the requirements of the Archeological Resources Protection Act of 1979, under which Chatters did his initial collecting. Without so much as mentioning Chatters's name, McManamon makes his opinion of those efforts clear enough:

> Knowledgeable and detailed inspection is needed.... The initial description and analysis [were] very brief and the existing written description is inadequate. Existing documentation about the recovery actions, items and remains discovered, location of the recoveries, and initial examination of the remains

consists of 13 pages of hand-written, difficult to interpret, and incomplete field notes. There is no map of the recovery site showing, even approximately, where the various remains were recovered.

During the month following the discovery, a few basic osteological measurements and nonmetric observations were made, [but] the method and techniques used for this recording have not been described. For example, fragments of some of the bones and the skull apparently were glued together. How such reconstruction may have affected the measurements and interpretations cannot be determined due to the lack of description of the methods and techniques used.... This documentation is not adequate for the DOI or COE to proceed with making decisions in this matter.

A single sentence from an earlier draft of McManamon's report, deleted from the final version, sums up the author's description of Chatters's methods and conclusions: "Considering the limited recording and uncertain data, the breadth and scope of the interpretations that have been made concerning this set of human remains [are] astounding."[4]

The remainder of McManamon's report provides an outline of the procedures needed to fulfill the provisions of ARPA and NAGPRA: first, proper examination, measurement, recording, and analysis of the remains, following the most up-to-date and thorough practices current in the profession; second, minute comparison to other specimens of similar age in order to determine which North American population, ancient and/or modern, the Kennewick skeleton most nearly resembled.

In case all these efforts proved inadequate to enable scientists "to make a reasonable determination about whether or not the remains are 'Native American,' as defined for the purposes of implementing NAGPRA," McManamon ended his report with a brief list of further tests to help settle the matter, among them "dating of the remains using radiocarbon techniques ..., DNA extraction and analysis to assist in inferring ancestry ..., and stable isotope extraction and analysis," not to mention "other chemical, radiological or physical methods and techniques." It was admitted that such tests inevitably entail the destruction of some skeletal material—"typically only a few grams"—and that such destructive testing would be deeply antipathetic to some Native Americans. If destructive testing were to prove necessary, Mc-Manamon wrote, it would not take place "without further consultation with the tribes." But consultation or no consultation, there was no suggestion that the testing would not sooner or later take place.

And just who would be performing these examinations, this battery of tests? "The investigations will be carried out by appropriate government experts or experts hired by the government for this purpose." A broad and reasonable criterion, but one that boded ill for the plaintiffs who had spent so much time and money pursuing a chance to examine the remains themselves. To be "appropriate," presumably one should be "unbiased"; and however scientifically qualified they might be to perform the examinations and tests envisioned, could plaintiffs to a suit against the government resonably be considered unbiased? Two years and half a million dollars into the case, the plaintiffs found that they had won their point and yet, by doing so, had disqualified themselves from enjoying the fruits of their victory.

Competition for Clovis? *11*

*J*ulie Stein, curator of archeology for the Thomas Burke Museum at the University of Washington, had worked long and hard as
local co-chair to make the sixty-
third annual meeting of the Society
for American Anthropology a success, but she was concerned that a
visiting journalist planning to attend
not come with any false expectations. "You may be disappointed," she
said bluntly. "Keep in mind what kind of group this is. Archeologists
spend a lot of their time alone in remote places with nobody to talk to,
so when they do get together, it's more like a high school reunion than
your typical scientific meeting."

Shortly after 8 A.M. on Thursday, March 26, 1998, the first full day
of the convention, Stein's warning seemed justified. Over two dozen
symposia on everything from excavating a nineteenth-century sugar
plantation in Panama to warrior maidens in pre-Christian Europe
were already under way on the sixth floor of the Washington State
Convention Center in Seattle, but most of the more than 3000 attendees seemed to be out in the corridor in a solid hand-shaking, back-
slapping mass, their bellows of greeting reminiscent of a seal beach in
calving season.

But a day or two into the five-day session, even an outsider could
see beyond the hey-hi-how-the-hell-are-you; could spot the elbow-

clutching hand and shifty, crowd-scanning eye of the job seeker; could spot the flushed face and overly loud voice of the careerist first-naming a buttonholed celebrity under the envious glares of less successful bounty hunters; could even assess the rise or fall of a speaker's reputation through the flow or ebb of the audience before the presentation.

One could also sense, behind all the incessant maneuvering, a continuous throb of special excitement. North American archeology is not, like physics or genetics, constantly rejuvenated by conceptual breakthroughs, but recently, under pressure from an accumulation of separate discoveries in the course of the 1990s, the ruling paradigm of the preceding seventy years, the Clovis-first paradigm, had begun visibly to crumble. The dust hadn't settled in March 1998, but the air was clear enough to show a profession in ferment, with dogmas under assault or in retreat, and the ozone tang of heresy everywhere.

By rights, Jim Chatters should have been having a grand time. When the SAA met the spring before in Nashville, Chatters had been facing an order to appear before the U.S. Attorney in Yakima with all of his notes and records concerning Kennewick Man. A friend on the Colville Indian Reservation had warned him to watch his back, that Jeff Van Pelt of the Umatilla wanted him prosecuted for (mis)handling the Kennewick remains; and his coroner pal Floyd Johnson had called after an interview with a Justice Department lawyer to say, "They're after your butt, buddy." A Corps of Engineers lawyer had been "extremely cold" to him, and other Corps employees had refused to meet his eye or shake hands with him in public. All in all, Chatters had been upset enough to claim his Fifth Amendment right not to incriminate himself in order to avoid providing evidence to Justice.[1]

A year later, as the SAA convened in Seattle, the sun seemed to have come out for Chatters. The threat of prosecution apparently had evaporated. For the first time in his life, his name and work would be as well known to the celebrated leaders of the profession in attendance as their names had long been to him. And he was scheduled to make a presentation on Kennewick Man at a Thursday morning seminar devoted to one of the hottest topics of the day: "Pre-Clovis Human Occupation of the Americas."

As it turned out, SAA 98 didn't prove quite such a supportive platform. Many of the field's greatest luminaries were in attendance, among them Kennewick Man plaintiffs Gentry Steele, Dennis Stanford, and Vance Haynes, but they showed no inclination to take a personal interest in the career of the now celebrated but controversial colleague who had precipitated the case in the first place. Oregon State University's Robson Bonnichsen, the title plaintiff in the suit, was more inclined to take Chatters under his professional wing, but he was preoccupied much of the time with cultivating the company of the person reputed to be the most important financial backer of his Institute for the Study of the First Americans: Jean Auel, author of the best-selling *Clan of the Cave Bear* novels, about an orphaned blonde Cro-Magnon adopted by Neanderthals.

The session at which Chatters spoke drew a good house, but curiosity seemed as much a reason for attending as anticipation of exciting research results. Its organizers, a self-funded California freelancer and the Department of Agriculture's cultural resource man for the western Midwest, were not among the top or even the second tier in professional reputation, and the papers presented by Chatters's ten co-speakers—among them a catalog of Native American tales of ancestral migrations, a survey of mammalian lifestyles on both sides of the Ice Age Bering Strait, and a rambling tribute to Franz Boas's insights into the links between human populations in the same region—were not the sort likely to erode the Clovis paradigm significantly. The most distinguished name on the bill, Anna C. Roosevelt of the archeology staff of Chicago's Field Museum (and the person who requested that Chatters be included in the line-up) chose at the last minute not to present her own paper, turning the job over to one of her students.

Chatters offered no new information or interpretations in his presentation. To the accompaniment of slides, he recapitulated the story of the discovery, repeated his conviction that the remains exhibited clear "Caucasoid" characteristics, but cited no numerical or statistical support for the statement, and concluded with a dramatic photo of Tom McClelland's portrait bust of K-Man-in-the-flesh lit from below ("our Kennewick Man-by-firelight shot") to convey a feeling, Chatters said, of sitting around the campfire with the young folk of the tribe, held spellbound by the stories the old chap had to tell. He would have stories to tell *us* too, Chatter said in closing, if only he and others were permitted to tease them out.

Leaving the hall for lunch, many in the audience were saying that the preconference buzz had been right on the money: Today's program had been a sideshow to satisfy the professional fringe: For what was *really* new on Paleo-Indian America, the don't-miss symposium was tomorrow morning's "What Happened 11,000 Years Ago in North America?" featuring papers by heavy hitters like Gentry Steele and Joe Powell, Vance Haynes, Don Grayson, and Gary Haynes.

As expected, some chewy and suggestive material was presented Friday morning, but auditors got an unexpected bonus when one of the "discussants"—distinguished guests appointed to provide a sort of postgame expert wrap-up for each symposium—decided to depart entirely from the material just presented to raise a more general issue. Lawrence Straus of the University of New Mexico began by reminding the audience that his own area of special interest was the European Mesolithic—the period of greatest flourishing of our relatives and perhaps ancestors the Neanderthals—and that he therefore had no professional axe to grind in expressing deep concern about the involvement of "some of our field's more well-known practitioners" in the furor over Kennewick Man.

As a specialist in the Europe of 40,000 to 20,000 years ago, Straus was particularly troubled by the "white wanderer" element in the evolving Kennewick Man myth as hinted at in Douglas Preston's *New Yorker* piece and even in scenarios spun out by established professionals with tongues sufficiently loosened by the heady draught of public interest. "We know that the fossil record shows a great deal of variability among early populations of the Americas and that there are quite a few alternative scenarios that can reasonably be put forward for who the ancestors of today's Native Americans were.

"But *all* these scenarios point to Asia as the point of origin, and to encourage the public to draw completely erroneous conclusions about the racial affiliation of Kennewick Man by referring to him as 'Caucasoid,' and speculate about a settlement route from Europe to North America at a time when there was nothing to be seen north of the Loire valley but solid glacial ice, seems to me to be not only unscientific but the height of irresponsibility." Sitting in the front row of the hall, in the direct line of fire of Straus's denunciations, was Jim Chatters.

Straus's strictures went unseconded, but unchallenged as well, an indication that the Kennewick story per se had little power to move an overwhelmingly professional audience. There had been little significant change in the facts of the case for more than a year now; the withdrawal from its control by the Corps' accident-prone handlers and the growing domination of its course by the Department of the Interior's understated, relentlessly professional team was beginning to choke off the controversial oxygen which had kept the blaze alive.

But the main thing cutting the Kennewick discovery down to size was growing competition from other recent and brand new discoveries of equal and, often, far greater scientific import. And the man responsible for the most significant of all had been sitting beside Straus at the Friday morning symposium: Tom Dillehay of the University of Kentucky.

Unlike many of his fellow discussants, Dillehay was not in town to present a paper himself. He was far from being under any professional pressure to do so. Indeed, for him, SAA 98 was little more than a victory lap. For nearly fifteen years, since the first publication on his research at a site in Chile known as Monte Verde, Dillehay had been the object of the professional scorn and abuse reserved for those who seriously threaten the conventional wisdom. But the year before the SAA gathered in Seattle, his work had received the kind of vindication that rarely comes to such an outsider. A dozen powerful champions of the standard Clovis model for the first settlement of the Americas had publically, though with varying degrees of enthusiasm, acknowledged that Dillehay's researches in a Chilean peat bog had blown their preferred paradigm sky-high. Before the first trace of the mighty mammoth hunters in North America, human beings had been stalking the creatures nearly 10,000 miles farther south. Clovis as a theory, was dead.

*I*n 1976, when he was a fresh University of Texas Ph.D. working at the Universidad Austral in Valdivia, Chile, Dillehay was shown some mastodon bones discovered along a logging road in Chile's temperate rain forests to the south. Dillehay thought he saw enough evidence of human agency on the bones—scrape marks where meat was removed, splitting to extract marrow—to make it worthwhile to take a look at the place where the bones had been found.

A first, cursory inspection of the site, the damp bank of a little tributary of the Rio Maullín called Chinchihuapi Creek, wasn't encouraging. There were bones a plenty, but none of the characteristic flakes of stone found in typical kill-sites such as Folsom and Clovis. And natural processes can produce cracks and breaks so like those made by human agency as to fool even an expert eye. Still, the site presented odd aspects. If this was merely an "elephants' graveyard," why were only rib bones to be found? And what, if not humans, had produced the numerous chunks of wood found in the sodden ground among the bones? Dillehay had little money and less encouragement from his superiors to continue, but continue he did. Had he known what was in store for him, he might well have chosen not to.

Ignored by the Chilean archeology establishment (both as a nonspecialist in coastal prehistory and as a *Norteamericano*), Dillehay got some of his new colleagues at the University of Kentucky to take part in excavating and/or analyzing the materials at the Monte Verde site, and he persuaded the National Geographic Society to provide some financing. In February 1979 (high austral summer), after two short survey seasons on the site and before serious excavation to expose the bone-bearing layer was complete, Dillehay and his diggers had a visitor: a retired curator of South American archeological materials at the American Museum of Natural History in New York.

Dr. Junius Bird was perhaps not the ideal person for National Geographic to ask to assess Dillehay's Monte Verde work. Though he had spent some time in southern Chile in the mid-1930s, Bird had earned his reputation in the 1940s and 1950s analyzing the textile industries of the prehistoric peoples of northern Peru. But it had been difficult to find anybody both professionally at liberty and willing to undertake the long journey. At seventy-two years of age (seven years after his re-tirement from the AMNH and three years before his death), Bird was not willing to spend more than two days at the dig site, and he devoted all of forty-five minutes to examining the artifacts and bones already turned up. Upon his return to civilization, first in Valdivia and Santi-ago and later in New York, Dr. Bird informed anyone who cared to lis-ten that in his opinion, there was no evidence that Monte Verde was a site worth digging.

Dr. Bird's dismissal was enough to injure Dillehay's credibility within the profession. But Dillehay's own findings did even more dam-age. The first radiocarbon dates obtained from charcoal at the Monte Verde site indicated that it had been deposited there a good 12,000 years ago. And as everybody "knew," the earliest reliable date of human occupation in the Americas was nearly a thousand years later than that—and 8000 miles to the north to boot.

In the face of almost universal disapproval and suspicion, Dillehay, with a stubborness and spunk worthy of the great nineteenth-century obsessives who founded the science of archeology, kept working Monte Verde through seven summers of relentlessly meticulous exca-vation, raising most of the money he needed to keep going from the University of Kentucky, persuading geologist, microbiologist, pathol-ogist, and agronomist colleagues there and elsewhere to apply their expertise to the materials found in the Chilean bush.

He was not rewarded for his persistence. In the seven years of dig-ging, not one Chilean specialist visited the site to observe the work in progress. At professional conferences back home, Dillehay encoun-tered colleagues who refused to shake his hand and, in private, encour-aged their students to regard him as a charlatan or lunatic. When the Smithsonian Press was persuaded to publish the first of two massive volumes on Monte Verde in 1988, the tide against Dillehay began per-ceptibly to turn, but it was very nearly ten more years before his work received full, public, professional vindication.

It came in February 1997, shortly after the publication of the sec-ond and final volume on the research at Monte Verde. The agent was David Meltzer, the Southern Methodist University archeologist who

for years had played devil's advocate to proponents of pre-Clovis settlement of the Americas. Impressed by the quality of Dillehay's published materials and troubled by the continuing covert, poisonous at-
tacks on him by professional associates, Meltzer persuaded a Dallas
plutocrat to pay the cost of sending a "blue-ribbon" panel of archeologists to Chile to inspect the site personally and challenged some of the
most resolute of Dillehay doubters to fly down to see for themselves.

They returned, if not uniformly converted, at least convinced that
the preponderance of the evidence was on Dillehay's side. In a review
article published the following May, Meltzer called Dillehay's work at
Monte Verde "a milestone in American archeology" that effortlessly
cleared the evidentiary hurdles demanded of any site claiming to be
earlier than Clovis: "unambiguous artifacts or human skeletal remains
in unimpeachable geological and stratigraphic context, chronologi-
cally anchored by secure and reliable radiometric dates."[2]

Converted in little more than a year from quasi-outcast to super-
star, Dillehay behaved modestly and with exquisite courtesy throughout his Seattle visit, but he would have had to be superhuman not to
enjoy the discomfiture of his erstwhile enemies, particularly when invited to "discuss" the papers presented at the Friday session. At lunch-
eon after the symposium, Dillehay, conspicuous among the bolo-tied,
cowboy-booted crowds in a tweedy outfit that would not have dis-
graced an associate professor of romance languages, remained resolutely above the fray. But when congratulated for being the man who
at long last had kicked the door open to serious consideration of pre-
Clovis human occupation of the Americas, he accepted the accolade
only dubiously. "Yes, I guess we kicked the door open all right. The
problem now is what's coming through the door after us."

*I*n concentrating on the importance of Monte Verde's early date—
thirty radiocarbon readings clustered around 12,500 years before
the present—early reports if anything underplayed the site's enduring
importance for archeology and for our understanding of the deep human past. The discoveries at Folsom and Clovis taught archeologists

where to look and what to look for in their pursuit of evidence of the settlement of North America. Monte Verde directed attention toward a vast research terrain that had previously been undervalued or neglected entirely, in large measure because of the power of the Clovis macho-myth.

Monte Verde is a "wet site." Organic material—flesh, wood, bone—tends to decay in surface water as fast as, or faster than, it does when exposed to air or soil. The breakdown is due not to the water itself but to the oxygen dissolved in it—directly or, by fueling the operation of scavenging organisms such as worms and bacteria, indirectly.

Stagnant water, on the other hand, acts as a preservative: Once its own oxygen content is used up, it is an effective seal against further corruption of organic matter buried in the soil it saturates. Boggy soils, peat bogs in particular, have long attracted archeologists' attention: From their airless, highly acid embrace have emerged some of the best-preserved ancient human artifacts and remains discovered anywhere.

The most significant materials discovered at Monte Verde were all but imbedded in a water-saturated peat layer that, stratigraphic analysis showed, had accumulated rapidly because of a rising water table and then had been effectively sealed off from air, stream flow, and animal disturbance by later, sterile, sandy deposits. The stone artifacts clustered here and there in the area were archeologically interesting, but hardly revolutionary: basalt and quartz tools shaped by flaking and chipping, along with what appeared to be naturally fractured river pebbles picked up for brief casual use and then discarded.

But it was not these artifacts, or the familiar paleo-litter of mastodon bones discarded after butchery, or even the circles of heat-hardened earth giving evidence of ancient campfires that most distinguished Monte Verde. The site, though obviously not a permanent village, was no overnight or weekend campground. In one of the two areas most thoroughly excavated, a criss-cross of logs set into the sand of the ancient creekbank defined a suite of side-by-side "rooms," one spacious enough to enclose a hearth a good five feet in diameter. Fallen among them, slender tree limbs showed how the "walls" of the hut assemblage were supported; scraps of animal hide, undecayed after more than twelve millenia, showed what the walls were composed of.

Even more exciting were traces of vegetable matter that only an archeologist would notice or value: seeds, stalks, skins, leaves, shells, roots of plants native to the site but also some native to coastal areas ten miles away; chewed and discarded quids of a medicinal herb that now grows no nearer than a hundred miles to the north; spores of another plant used today by natives of the area as a kind of talcum to soothe skin irritations. The Monte Verdeans made use of plants of all seasons, which suggests that their settlement may have been a year-round base camp, something extremely rare among hunter–forager peoples of such ancient times.

Only when a site like Monte Verde comes along does one realize how much of our image of life in post–Ice Age America is sustained by a mere gossamer of speculation spun around precious few facts. Like the Clovis and Folsom peoples, the Monte Verdeans ate big game. But they also, and apparently a lot more often, hunted small game (with *bolas*, the stone-and-thong tanglefoot weapons still used by South Americans in historic times). They made tools of stone but also of bone, cane, and half a dozen different kinds of wood. They roasted frogs, gathered mussels, ground seeds into flour, harvested greens, cooked roots, poulticed their illnesses, laid out their camp with distinct work and dwelling areas, and built privies (or at least buried their wastes). If Dillehay's cautious interpretation is correct, one odd-shaped structure, set apart from the rest and littered with its own spectrum of remains, may have been a site for ritual observances—or perhaps just for getting high. The Monte Verdeans seem to have behaved, in other words, very much as our omnivorous, restless species has always behaved: They ate and used everything around them worth eating or using and picked their teeth with the leftovers.

Until the validity of the Monte Verde evidence was finally granted by the gurus of the trade, it was the early radiocarbon dates, naturally, that drew the most attention and provoked the most disagreement. Once those dates were confirmed, however, its implied portrait of un-heroic routine daily existence couldn't help but enhance the significance of similar evidence that has been around for years, in some cases for decades: ancient remains of a tough net woven from grass found high in a Rocky Mountain cave, just the right mesh size to entangle the mountain sheep that still inhabit the area; seed-grinding stones from caves occupied when the Bonneville Salt Flats were still a freshwater ocean, tools still used to crush corn for the daily tortillas in remote areas of Central America.

In the kind of domestic economy suggested by these traces, women (and even children) aren't condemned to sitting around waiting for the hunters to bring home the bacon. They're part of the action. In the 1990s, stories celebrating the role of women in hunter–gatherer society proliferated in popular-science magazines—accounts that reject old-fashioned caveman machismo for a new, feminist interpretation of prehistory.

But wait a minute. Isn't this story starting to sound awfully familiar? The epic, continent-conquering Clovis suited the unexamined world view of the scientific storytellers of post-frontier, imperial America. Is the new, downscaled, less macho, more domestic conception of early humankind in the Americas just a projection of the prejudices of the educated classes of our own day? Is it a nonviolent, egalitarianized world view just as myopic and misleading as what preceded it?

Up to a point, no doubt. But it isn't just our image of the Clovis lifestyle that new discoveries such as Monte Verde are forcing us to revise; it's the whole story of *how* the Americas came to be settled, not just when. After decades of stasis, archeology in the Americas is experiencing continuous ferment. Explanatory myths that have served for decades are being revised or discarded in the face of new evidence; dates long believed firm are crumbling, as old sites are re-examined in light of fresh data from others. The stolid institutional structure and old-boy network that has dominated archeology for more than a century is collapsing under pressure from other scientific disciplines, from social change, and from sheer economic constraints. It is an uncomfortable time for traditionalists. But for the young, the independent, and those willing to change, that discomfort is only an index of their own liberation.

The Bone Business 12

*M*onday, July 20, 1998, dawned unseasonably cool for Tucson, Arizona. A high overcast had followed the thunderstorms of the night before, and by 9 A.M. the temperature had barely grazed 80. It wouldn't hit 100 until mid-afternoon, and by that time the day's work would, with any luck, be over. For the purposes of a National Geographic or Discovery Channel archeology documentary, this narrow strip of drab sand, between the southbound Ina Road on-ramp to Interstate 10 and the chaos of a huge waste-water treatment plant construction site, was entirely lacking in visual zip. But the dig under way here, centering along a meter-wide trench deeper than the height of a tall man and protected from sun and rain by an incongruous string of tents resembling miniature wedding marquees, was as important as many in such more photogenic locations as the Anatolian interior, the Mayan jungles, and the Great Rift Valley of Africa.

In sober fact, the excavations at Las Capas are typical of a great many historically significant sites investigated since amateur antiquarianism began to give way to professional archeology. The direction of research for two hundred years has been determined at least as much by economic as by intellectual forces: by the decisions of railway builders in early industrial England and France, and in America today by the decisions of developers of high-rises, highways, and parking lots.

In 1993, when the Arizona Department of Transportation asked for bids for an archeological survey of the terrain along Interstates 10 and 19 west and south of Tucson, the history of how farming came to this part of the Sonoran desert was pretty clearly blocked in. Early hunter–foragers had left their traces in the lowlands hereabouts, but it was believed that New World agriculture first came to Sonora via the uplands, making its slow way up from central Mexico where it first appears in the archeological record, perhaps as much as 7000 years ago.

Because of the extreme seasonality and unpredictability of rainfall, not to mention the brutal heat of the desert summer, it was not until well into the Christian era—the dates suggested by various authorities range from A.D. 300 to 700—that Southwestern peoples managed to set up permanent settlements in the lowlands. These people, collectively called the Hohokam (from Pima Indian words meaning "all used up") developed remarkably sophisticated ways of controlling and directing the water from the few perennial rivers crossing the Sonoran desert to water their fields of maize, squash, and beans. When a series of catastrophic floods during the 1350s left them unable to rebuild their system of canals, the Hohokam disappeared as a living culture, leaving south-central Arizona dotted with the ruins of villages and full-fledged towns which still attract tourists today.

The standard explanation of how farming came to the Southwest is intimately associated with the name Emil Haury, the man who all but dominated the study of archaic Southwest civilization through the mid-twentieth century. Haury believed that the evidence showed agriculture arriving in the area from central Mexico via a highland route, moving with excruciating slowness from one well-watered niche to another until it reached the Southwest in upland Anasazi country, and only then moving down into the less temperate lowlands.

By 1993, when the first contracts were awarded for the Interstate 10 survey along the east bank of the Santa Cruz, it was already believed that agriculture had "come down out of the hills" a good deal earlier. Still, scientists working for the winning bidder, Desert Archeology, Inc.

(DAI), were not expecting to upend the textbook account of Southwest civilization with their dig. But the surprises started almost as soon as the work began, at the site of the main freeway exchange for North Tucson off the drab strip development misnamed the Miracle Mile.

Here, where the Santa Cruz channel takes a sudden narrow swoop to the east in its generally southerly course, crews cutting survey trenches under archeologist Jonathan Mabry kept turning up traces of human occupation a meter or so down in what was supposed to be sterile alluvium. At the base of this layer lay circular traces of so-called pit structures—some for storage, some clearly the foundations of dwellings of some sort, all ringed with postholes that once contained willow and cottonwood poles to support roofing of grass thatch.

Sufficiently intrigued to invest time and effort in a wider excavation, Mabry and his workers brought in loaders and backhoes to strip away the overburden from a substantial area. What they found shocked even them: traces of 176 structures, including some apparently intended for communal use, tightly grouped together over an area of less than two and a half acres.

Over the next four years, seven freeway-side sites including Las Capas were excavated by Desert Archeology. With each the pattern became clearer. Far from being temporary in any sense, the settlements along the lip of the Santa Cruz floodplain were, if not full-fledged farming villages, the next thing to it. At a period when the Santa Cruz appears to have been a more or less permanent stream, the inhabitants of the area channeled its water through a complex system of canals and ditches to irrigate their crops: not just maize, squash, and beans but cotton and tobacco as well—the first occurrence of the latter in the Southwest.

The sites varied in richness of surviving material, but as a whole they presented a picture of a fully developed irrigated-farming peasant society as known from other digs around the world and in surviving communities in historic times. The inhabitants ceremonially buried their dead, both within the confines of the community and in a formal "cemetery" plot: a practice considered a dead giveaway of adaptation to sedentary lifeways, a form of laying ultimate claim to the land.

They traded, and not just for precious tool obsidian from distant parts of the Southwest, but also for ornamental shells, some from the Sea of Cortez near the mouth of the Colorado, others from the Pacific near modern San Diego a good 500 miles away. There is evidence that they used the bow and arrow in hunting; it is certain that they made and used pottery—not a practical technology for nomadic peoples. And they did it all at a date somewhere between 800 and 1500 years earlier than they were supposed to have been doing such things in the Arizona lowlands.

Once again, archeological experience had ossified into archeological dogma, to the detriment of knowledge. Southwest archeologists were used to surveying sites by putting on their boots and walking. Thanks to the arid climate, they found artifacts almost everywhere they walked. In areas where they didn't find artifacts, they tended to assume that the artifacts weren't there. But they were, and only three feet beneath the surface, though it took a freeway ramp enlargement project to turn them up. The textbooks were wrong. Time to start over with a clean slate.

At luck would have it, some of the Tucson digs were in the hands of a man with hands-on experience collecting the kind of data they presented. Jonathan Mabry made his first small contributions to archeology as a teenager, illustrating artifacts for excavators at the University of Kentucky. In college he spent his summers on digs in the Middle East, and he ended up writing his thesis on early agricultural settlements in the Jordan Rift Valley. He had spent a year there walking down wadis—the Mideast equivalent of arroyos—and reversing classic archeological method. His unconventional approach was to look for buried prehistoric sites and compare the potsherds and stone tools found in them to similar materials of known date to establish the geologic sequences in which they had been buried, in order to reconstruct what was happening with the landscape over time.

He found a very definite pattern. When traces of occupation by agriculturalists near a streambed appeared, they did so at times when, and at locations where, the stream was aggrading—that is, depositing more sediment from upstream than it was washing away downstream.

The native ancestors of modern grain crops flourish under just such conditions, in well-watered rich alluvium. As climatic conditions brought these soils farther down the wadis from the higher ground where such grasses grew in the wild, the grasses moved downstream with them—and with them, apparently, their harvesters.

"Even hunter–gatherers—no, *especially* hunter–gatherers—are extremely observant of the landscape," Mabry says. "When the wild wheat and barley followed the wadis downstream and spread across the alluvial fans where the hills met the plain, they realized, 'Hey, we can enhance this!' They were doing flood-farming at the mouths of the wadis, planting their wheat and barley on those alluvial fans and diverting that runoff through channels into their fields. The earliest type of agriculture in the world. And not all that different from what we see right here, where they were planting maize on land flooded by the river, expanding the cultivable area with canals."

*M*abry and Co.'s discoveries in the Santa Cruz Basin are forcing North American archeologists to make some radical revisions in their timetables, but so far they provide no reason to throw the timetables away entirely. The crops and technologies of farming may have arrived in the Southwest a thousand or even two thousand years sooner than formerly believed, but they still arrived from the south. The Tucson farmers may have been newcomers to the region, or they may have been long-established inhabitants who adopted a new lifestyle, but farming always follows hunter–gathering in the archeological sequence, never the other way around. The series of Southwest cultural traditions established by Alfred Kidder in the 1920s and clarified and elaborated by his successors ever since stands unquestioned.

A far different challenge faces the standard model of how the Americas came to be settled in the first place. As recently as 1995, the Clovis-first theory still dominated responsible discourse on the subject.[1] But with leaders of the profession accepting Monte Verde as a legitimate site as old as, or older, than Clovis, and with the artifacts discovered there bearing only passing and possibly fortuitous resemblance to the highly individual Clovis style, archeologists have been forced to reconsider a lot of previously ignored or underrated bits of evidence suggesting that the mighty mammoth hunters weren't the only folk on the North American scene as the Ice Age waned—and that they may even have been Johnny-come-latelies.

The Clovis-first paradigm suffered from the beginning from a serious weakness. In a way, it was founded not on evidence, but on lack of evidence, even after precise methods of dating archeological sites came along. It was known that human beings were hunting mammoth on the shores of Lake Baikal in eastern Siberia 25,000 years ago. And 10,000 years later a similar folk were hunting the same prey along the banks of the Aidan River, over a thousand miles farther east and north. Between 13,000 and 12,000 years before the present, a land bridge existed between Siberia and Alaska. Very soon after the latter date, mammoth hunters appear in North America. Granted, the dots of evidence describing this geotemporal trajectory lay many thousands of miles apart. But the pattern was so clear that additional confirmatory finds would surely turn up in time.

The trouble was, they didn't. When the first intrepid archeologists began to brave the hostile Alaskan environment, the scanty remains they found of early human occupation didn't *look* at all like the Clovis tool kit or much like the cruder Siberian materials, either. In particular, the long, fluted-stone projectile point that defines the whole Clovis tool tradition was absent. Worse yet, when accurate radiocarbon dates for the northernmost occurrences of Clovis materials were obtained, they uniformly tended to be *younger* than materials recovered from Clovis sites farther south.

The route supposedly followed by the mammoth hunters, straight from interior Alaska to the Great Plains, began to show signs of strain as soon as evidence began to replace extrapolation. Unlike the Clovisites, Canadian geologists looking at the terrain exposed during the recession of the Ice Age glaciers were not prepared to vouch for an "ice-free corridor" opening up from north to south in time for the hunters to keep their appointment with archeological destiny. Nor were their climatologist colleagues more sanguine. Even if a corridor of sorts developed, they said, a narrow, wind-scoured canyon between cliffs of ice, its floor paved with sterile boulders and sand and drowned by frigid meltwater lakes, would not seem likely to have been able to support the big game the hunters supposedly lived on, or any game at all, for that matter.

At this point, the sensible course would have been to label the Corridor–Clovis account a plausible, but far from watertight, theory and to welcome alternative interpretations of the data. Instead C–C hardened into dogma, and anyone who proposed an alternative was more or less publically labeled an outsider. This stigma attached particularly to theorists who didn't much care *who* settled America first, but instead tried to figure out *how*. If the ice-free corridor was starting to seem an unlikely settlement route, they asked, what was the alternative? Why, the coast, of course. For most of human history, our species has clustered near the ocean, where the climate is more temperate, food more abundant, and transport easier than in relatively inhospitable mid-continental areas.

Almost from the day it was broached as a possible alternative to the canonical ice-free corridor, advocating "the coastal route" has been a sure way to raise eyebrows, rouse gibes, and lower one's chances of gaining admittance to the inner circles of North American archeology. The arguments against human beings entering North America via the Pacific coast were simple and, to their adherents, overwhelmingly convincing. Until 10,000 years or so ago, the coastline of North America from the Gulf of Alaska to Vancouver Island was little more than a wall of ice, they said, every present-day river valley the channel of a glacier plunging out to sea. Where, pray, was there dry land for the unknowing invaders to tread upon, let alone food to sustain them on their journey?

The first scientist to attempt a comprehensive answer to those questions was Knut Fladmark, an archeologist with British Columbia's Simon Fraser University. In a long, cautiously argued paper published in 1979,[2] Fladmark suggested that, given the sheer complexity of the terrain in question, was it likely that *every* mile of coastline from Anchorage to Coquitlam was continuously icebound? Clearly a great many problems facing the mid-continent ice-free corridor idea had been underappreciated.

In historic times, native peoples of the Alaskan coast managed to coexist successfully with glaciers calving icebergs from the very valley mouths. Was it not at least possible that along 1200 miles of shore, patches of land unscoured by ice existed, where the newcomers might find sufficient plant and animal food to survive? The idea didn't seem any more unlikely than people traversing an equal distance down a dismal cranny between two ice caps. As for the undoubted icy patches between "refugia," perhaps watercraft....

Nonsense, said the traditionalists. Everyone knew that the building of boats, even primitive boats, comes relatively late in the human adventure. For that matter, although the reason remains obscure, it is well known that even in regions where fish and other marine foodstuffs were easily available, early humankind failed to take advantage of them.

The most powerful argument against the very idea of a coastal route into the Americas, however, was a kind of anthropological reductio ad absurdum. Suppose that there *had* been a migration along the coastline, went the argument. Whether with or without the assistance of watercraft, it must have followed closely the coastline of the time. But the coastline of the time lay anywhere from 10 to 150 miles westward of today's and is drowned beneath over a hundred feet of ocean, because the world's sea level has risen through the melting of the ice caps. Evidence of human passage, if any, is irretrievable. Thus, even if the coastal hypothesis were true, it will never be affirmed, so the whole idea is as far beyond the realm of scientific study as life in the fifth dimension.

So confidently and often has this argument been put forward that even people who realize its weaknesses prefer not to engage with its proponents. As a matter of fact, there are a number of areas along the Pacific coast of Canada where the Ice Age coastline is *not* submerged but is actually higher in elevation than the present-day coast. What the preceding argument omits is the phenomenon, often overlooked by nongeologists, called isostacy. The crust of the earth is fairly rigid, but beneath the crust lies another layer that, though it is not exactly fluid, is sufficiently pliable to deform under pressure. Piling a half-mile or so of solid ice on its surface makes the underlying crust ride lower in its viscous underpinning. Remove the additional weight, and the crust—though very slowly—is once more buoyed up.

The dynamics of the Northeast Pacific coastline during the last 100,000 years or so of intermittent glaciation are only beginning to be traced in detail. It is clear, though, that however hostile the environs were to humankind 12,000 years ago and (thanks to foul weather, jungle-thick vegetation, and large predatory animals) to archeologists today, opportunities abound to test the coastal hypothesis. The problem, until recently, was a lack of will and funding to do so.

Even before the first hard evidence of ancient human presence along the North American coast turned up, some of the traditionalists' objections to the very notion were beginning to falter. The idea that watercraft were unknown to late Ice Age peoples began to buckle severely after 1964, as radiocarbon dating pushed the likely first date of human arrival on the Australian continent back to 40,000, then 50,000, perhaps even 60,000 years ago. Could occasional incautious unfortunates blown across the Timor Sea on a floating log really have led, in time, to an enduring breeding population?

Even more intriguing were 1962 discoveries on Anangula Island in the eastern Aleutians, where remains dating to at least 8500 years ago gave tantalizing hints of "maritime adaptation," including whalebone. Traces of recurrent human settlement going back perhaps as much as 9000 years were found near the immemorial seasonal salmon-fishing site at Ugashik Narrows. On Kodiak Island and on Cook Inlet west of today's Anchorage, further traces of coastal occupation before 7500 B.C. kept turning up. But by 7500 B.C., of course, world sea levels were back to contemporary levels. Such traces proved nothing about earlier occupation of the coast.

Having spoken his piece, Fladmark more or less retired from the controversy to concentrate on other matters, but his ideas continued to simmer among younger and more open-minded students of the subject. In the twenty years following the publication of his paper, numerous coastal sites of apparently very early date turned up in the literature, but none under circumstances that constituted a direct challenge to the exclusive dominance of the Clovis paradigm. As for the undersea sites that might support Fladmark's hypothesis, they remained undersea, inaccessible to fan and foe alike.

Among the biggest aftershocks of Monte Verde's transformation from object of derision to paradigm buster, none was more satisfying to the heretics than that produced by a four-inch triangle of basalt sluiced out of muck dredged from the ocean bottom half a mile off the coast of British Columbia's Queen Charlotte Islands: a scraper or knife clearly chipped into shape by human hands.

The discovery, made on May 17, 1998, came about by anything but chance. Every summer for four years, Parks Canada archeologist Daryl Fedje and the Geological Survey of Canada's Heiner Josenhans had been taking advantage of every day of decent weather in these treacherous waters to assemble a sonar map of the sea floor, which still bears traces of the hillocks, riverbeds, and estuaries that lay upon it when sea level was two hundred feet lower than it is today. Drawing on that body of geographic information and using the global satellite positioning system to orient themselves precisely above locations likely to have harbored human occupation, Fedje's team spent day after backbreaking fourteen-hour day dredging. Their payoff was one crude artifact, but that one was enough to turn "the coastal hypothesis" from hypothesis to fact.

*D*AI's research along the I10 corridor is still not complete, but several fat volumes containing the results of six years of completed work have already been published. This is nearly without precedent in a field where it can take decades for the full record and data of important digs to make it into print, and where many finds of crucial importance—the 10,000-year-old bones and artifacts from the drowned Marmes site are an example—don't ever take their rightful place in the evolving narrative because they were never properly published at all.

Contract archeology, long scorned by scientists with academic appointments (or at best seen as a way of bringing in a little money to keep grad students fed and occupied) has grown so important that it threatens to become the well-funded tail on a severely emaciated dog. When a University of Arizona graduate named William Dolle founded DAI in the mid-1980s, contract archeology was a rapidly growing field, thanks to proliferating local, state, and federal laws requiring archeological surveys of any terrain proposed for development. With academic jobs hard to come by, a lot of fresh archeology Ph.D's founded independent companies to meet the demand: too many for the market to absorb, as it turned out. Most government contracts are put out to competitive bid, and in most archeology programs, courses in cost–benefit analysis and personnel management aren't a requirement. Thus, a lot of hopeful rec-room survey-and-salvage operations soon put each other out of business through underbidding in their eagerness to score a contract.

Dolle's firm survived to become a leader in the new field, one of a dozen or so specializing in the issues addressed and the techniques required in a particular archeo-cultural region of the country, and it earned a national reputation through the quality of its work—and its publications. Mabry, the man in charge of several of DAI's freeway-side digs, is already a second-generation contract hand. Since taking his Ph.D. degree in 1992, Mabry has done no other kind of digging and has no regrets.

"When I got out of the University of Arizona the academic jobs were few and far between, so when I was offered a full-time position with Desert Archeology I grabbed it. I've never had reason to regret it. I have had the good fortune of digging one really interesting site after another. There's still a lot of tension on the part of academic archeologists about contract archeology, and that's understandable, because there is a growing recognition that contract is where it's at. We are doing by far the major part of the digging in this country year in and year out, we're the ones finding the data, at a very rapid rate, and most important we're the ones with the money to do the job right.

"Getting good data doesn't do much for science if it doesn't get published. For some reason, most of the granting sources for academic archeologists are willing to pay for the digging, but they don't understand that's just the beginning. For every month we spend in the field there's probably six months' worth of analysis and writeup. When we bid a project, all that's written right into the budget. But we're talking really about enormous differences in the kind of money available. Where land development is concerned, there's money around, and contract archeology is development-driven. An academic archeologist probably feels lucky to get $15,000 for a project. The one I'm working on here is budgeted over $2,000,000."

Not all academic archeologists face such extreme financial constraints. But very few have the freedom of SMU's David Meltzer, working under an endowment that both allows him to choose what site most merits investigation and also provides funds for the unglamorous but essential work-up time needed to turn raw data into publishable research. A few others have become sufficiently well known outside the profession to attract funding of their own: Donald C. Johansen's charisma and his early-hominid discoveries in East Africa's Great Rift Valley have attracted sufficient private backing (reputedly from the oil-rich Bass family) to allow him to set up his own anthropological think-tank, the Institute for Human Origins, and to pull up stakes and move the entire operation to the University of Arizona when he felt underappreciated at Cal Berkeley.

Robson Bonnichsen, lead plaintiff in the Kennewick Man case, is not so widely known outside the anthropology business as Johansen, but he too heads his own research organization, the Center for the Study of the First Americans. Once operating under the aegis of the University of Maine, Bonnichsen too hit the road with his CFSA when he found himself embroiled in conflicts with other Maine faculty. He settled in by invitation at Oregon State University, where he remains today, though not without continuing friction with his new colleagues. Again, CFSA has its own board of directors and does its own fund raising, so Bonnichsen has a good deal more latitude than traditional academic investigators in choosing research goals.

In the United States, academic archeology still enjoys the most prestige, but along with that produced under government contracts by the best contract archeology companies, most of the solid primary excavational work is being done directly for the government, by archeologists on the payroll of the Bureau of Land Management, Park Service, Corps, or other federal agency. To an increasing degree, Native American governments (such as those of the Umatilla, Colville, Nez Perce, Yakama, and Wanapum in the Pacific Northwest) are drawing on available government cultural resource funds to embark on their own programs, like Jeff Van Pelt's for the Umatilla.

Paradoxically, researchers working for perhaps the most prestigious name in American science are among the most financially constrained. The Smithsonian Institution's funding comes directly from Congress and is supposed to suffice for the research needs of the myriad subdivisions of the sprawling complex along the Mall. Smithsonian scientists are specifically forbidden to apply for grants from the biggest supporter of pure science in the land, the government's own National Science Foundation.

With more and more buildings to maintain and a payroll locked into the civil service system, most Smithsonian divisions are chronically short of money for research and yet, as an enormous government-underwritten operation, are awkwardly positioned to ask outsiders for help. Just as in the American not-for-profit arts sector, more and more of institutional scientists' time has to be spent looking for money in order to get any original work done at all. And, as in the arts, the choice of work to do is increasingly dictated by the publicity value of the work in question. One can imagine the frustration of Smithsonian plaintiffs Douglas Owsley and Dennis Stanford when the Kennewick remains, not only a potentially important scientific discovery but also a sure-fire public relations fund-raising gold mine, were snatched out of their control only days before they were due to take charge of those remains.

O wsley's frustration must have been particularly acute because Kennewick Man seemed to promise an opportunity to complement and enhance results he'd obtained from another human specimen only recently discovered to be of comparable age: the so-called Spirit Cave Mummy, first discovered in 1940 in a cave on the edge of the alkali flats east of the little Nevada crossroads town of Fallon.

These remains (not deliberately mummified but merely desiccated by the bone-dry air) ended up in the collections of the Nevada State Museum in Carson City, where they lay neglected in a basement amid thousands of other specimens, artifacts, and curios until the passage of NAGPRA in 1990 forced the collection's curators to resolve their status. Even then things moved slowly. Only in 1994 was material from the burial sent off to Ervin Taylor's radiocarbon dating lab in Riverside, California, and it was not until April 1996 that Taylor's results were announced: Spirit Cave Man had died over 9400 years before.

Spirit Cave Man instantly became the anti-repatriation campaign's poster child (until the discovery of Kennewick Man four months later knocked him from the front pages). The Paiute-Shoshone Tribe of the Fallon Indian Reservation near Spirit Cave had long been demanding repatriation of Nevada State Museum holdings of Native American materials. But, museum authorities maintained, if they had promptly repatriated the mummy, its immense scientific importance would never have been established. The argument might have been more convincing had the remains been more thoroughly studied during the more than fifty years they had been gathering dust until NAGPRA focused attention on them, or even during the forty years since radiocarbon dating became practical.

As with Kennewick Man, public and media attention was directed primarily to the mummy's presumed non–Native American racial characteristics. In an official museum statement, curators Donald Tuohy and Amy Dansie refer to bone expert Gentry Steele's analysis of the physical characteristics of the Spirit Cave skull as "consistent with other early Americans reported by him, and different from modern Indians" to reinforce their own judgment that "the grave associations tied to these ancient dates do not show any demonstrable relationships to the historic material culture of any living Native Americans." Indeed, they conclude, "as far as the scientific data reveal so far, there are no direct descendants of the early Americans living today"—strong language to use in advancing a thesis that would be impossible to prove no matter how much evidence were to become available.

It is in those "grave associations" of Spirit Cave Man that his real significance lies. Although the mummy was mentioned in a short museum report in 1969, the clothing on the body and the reed matting in which it was wrapped for interment remained unstudied. When a close examination finally took place, it was discovered that the weaving of the tule reeds composing the matting was remarkably complex, a form of diamond-pleating difficult to achieve with any precision without the use of a loom.

The robe of twisted strips of skin and the moccasins pleated from three different kinds of animal hide also testify to a highly complex fabric technology—at a time when inhabitants of the Great Basin were thought to have been barely more advanced technologically than their mammoth-hunting ancestors. All the evidence was there to see and had been there to see since 1940. No one had bothered to look until NAGPRA forced them to.

Like most of the nearly eight hundred sets of human remains in the museum's warehouses, the Spirit Cave mummy is not the "property" of the state of Nevada, but is held in trust for another agency—in this case, the Federal Bureau of Land Management (BLM). Nevertheless, museum staff were left to interpret the provisions of NAGPRA pretty much their own way until their announcement of the ancient provenance of the skeleton raised the political stakes.

By late 1997, staff had been persuaded that if they wanted permission to continue studying the remains, they had better make some gesture to the Native Americans who were lobbying the BLM for immediate and unconditional repatriation. A meeting was called by the Museum Anthropology Department at the Pyramid Lake Tribal Council Meeting Hall in Nixon to give members of the Paiute tribes of the Lahontan Basin an opportunity to express, to anthropologists from both the museum and the BLM, their feelings about archeology and repatriation.

Judging by the report published months later in a museum newsletter,[4] the Native Americans took full advantage of the opportunity.

"After the first three minutes had gone by," wrote Dansie and Tuohy, "it was clear to the State of Nevada and federal employees that we had driven into a trap, à la Major Ormsby of the Carson City Rangers who lost virtually all of his men and his own life, too, during the first battle with the Paiutes during the Pyramid Lake War of 1860."

Fortunately, no lives were lost in the engagement of November 25, 1997, though some tempers obviously were, on both sides. The anthropologists pointed out that "from present scientific knowledge ... a large percentage of the burials from the Lahontan Basin were 'unaffiliated'" and reminded the Native Americans present that a handout stating this had been distributed at an Intertribal Conference in Nevada only five years before. But it was all to no avail. Instead of acquiescence, "what all three of us got from the meeting was a history of what the Anglos did to the Indians for the last 180 years (in both Paiute and English), at times bordering on pure rudeness! ... Eventually, the United States soldiers conquered the Northern Paiutes and won the war of 1860—or did we?" The choice of pronoun says it all.

Three Years and Counting... 13

*E*very reporter knows that nothing takes the fun out of a good
story like a surplus of hard facts. All
the sharp outlines that looked so
clear, all the issues that stood out so
prominently, soften and blur when
they are illuminated from too many
sides at once. Drama, in journalism
as on stage, depends on contrast,
and contrast is hard to achieve when you're up to your neck in detail.

On paper, Frank McManamon's plan for a measured, deliberate,
thorough scientific examination of the Kennewick remains looked like
the most obvious, pedestrian approach imaginable. In a case in which
the few facts to be established had long ago been swamped in speculation and innuendo, such an approach could be perceived as a plot most
Machiavellian, and so it was taken by the plaintiffs in the case.

At a May 1998 hearing called by Judge Jelderks to learn how the
parties proposed to work together toward a resolution, the plaintiffs'
lead lawyer Paula Barran instead raised the temperature with a fusil-
lade of charges about federal mishandling of the bones while they were
in storage at Battelle Institute. Not only had there been a "theft" of
bones from the Battelle locker in April, Barran claimed, but only two
days before the hearing, she had been informed that "an employee of
the Battelle Institute, Mr. Darby Stapp, who has a significant position
in the Cultural Resources Program, is actually married to the archeol-

171

ogist for the Umatilla.... Mr. Stapp is one of the custodians of keys to the vault room.... It is also a fact that Mr. Stapp ... and his spouse have published articles which are highly critical of the scientists' positions in this lawsuit."

The source of this information, according to Barran, was Dr. Jim Chatters. Dr. Chatters was also the source of Barran's next example of government bungling: the apparent disappearance of two or more scientifically important bones from the collection, seemingly while in government custody. "They were present when Dr. Chatters had the bones, they were present when he videotaped the packing, they were present when he turned them over to the Benton County coroner.... And it would not be until March 1998 that the Government came back to [the judge] and reported that these bones were missing."

All in all, said Barran, there was ample evidence that the physical safety of the remains could not be guaranteed in their current repository, and the plaintiffs therefore proposed "that from this point forward the Kennewick Man skeleton be under the control of a neutral who answers to the Court because the Government has shown, despite a very long leash from this Court, that they are irresponsible in protecting these bones." She also twice mentioned a suitably "neutral" party to take over curation: the Smithsonian Institution, where two of the eight plaintiffs were employed.

The government attorney present that day, Robin Michael, did not feel it necessary even to rebut the notion that the Smithsonian Division of Anthropology could be considered a neutral repository for the remains. She concentrated on disentangling fact from innuendo in the plaintiffs' arguments, on introducing Corps archeologist Michael Trimble, and on establishing his expertise in the proper handling of archeological remains. She also elicited his opinion that the Kennewick remains would indeed be better housed in a facility equipped for professional curation and that the Thomas Burke Museum of Washington State in Seattle was a suitable location, being accustomed to handling paleontological materials and being neutral ground between the contesting parties.

In earlier statements, Judge Jelderks had frequently expressed displeasure with the Corps and its lawyers for their failure to provide the plaintiffs and the Court with timely and complete information when asked to do so. In his summing up of what had been accomplished during the hearing of May 27, he for the first time began to manifest an awareness that the plaintiffs' arguments might be driven by interests beyond the purely scientific. While granting the Smithsonian's Douglas Owsley and Dr. Chatters an opportunity to confirm with their own eyes the conditions under which the remains were being held, he almost pled with the parties "to have a good discussion among yourselves knowing that the remains are going to have to be moved to see if you can agree on a facility...." But, he continued,

> I don't have a high degree of confidence that you can. One of the interesting things to me has been the real adversarial nature of these proceedings from when the case was first filed on what really I would expect to be not quite so adversarial. It's not like a divorce where people are highly emotionally involved or somebody wants to get somebody else's money and the party that is being sued doesn't think they're entitled to it....
>
> The scientists appear to be well motivated.... What hidden motivation might they have to stir up this Court proceeding or to create these issues for the defendant if their motives aren't pure and their intentions good...?

Without answering his own question, the judge gave the parties a month to come to an agreement on a neutral site for curation if they did not want him to make the decision for them.

Instead of negotiating, the plaintiffs spent the ensuring month of June attacking the credibility of the government's suggested repository. Because independent consultants sent to evaluate several potential curation sites had given the one in Seattle as good a rating on facility and scientific grounds as others more distant, their attack focused on the neutrality issue instead. To demonstrate that the Burke Museum was by no means a neutral scientific bystander, they produced an e-mail reply by a Burke staff scientist to plaintiff Richard Jantz's September 1996 call, addressed to hundreds of his fellow professionals, for support for his position. Although the plaintiffs asked the court not to make the text public—"to avoid embarrassing the author," it was said—they did describe it publically as "very offensive" and "replete with insult, misinformation and accusation." If that wasn't enough to establish the unsuitability of the Burke Museum to serve as a neutral repository, there was also the damning fact that "on June 2, 1998, the Museum Director [archeologist Karl Hutterer] told [plaintiff] Dr. Robson Bonnichsen that he thought the whole affair could have been avoided if the plaintiffs had fairly contacted the tribes at the beginning of the controversy." Despite this plain evidence of prejudice, however, the plaintiffs were still prepared to allow the government to specify the Burke as the K-Man's new home—*if*, in return, "in the event that plaintiffs have any complaints about the Burke Museum staff's treatment of plaintiffs or public or private statements of bias about this case, the remains will be transferred forthwith to the Smithsonian Museum Support Center."

Despite this generous offer, the feds stubbornly continued, with the judge's bemused acquiescence, to plod toward a transfer of the remains from Richland to Seattle. Finally, after summer had faded into fall, the handover at last took place. On October 29, after a marathon session in Richland at which the Smithsonian's bone man Owsley inventoried and examined all the material bagged up by Chatters over two years before, the bones, packed in two snap-top Rubbermaid storage tubs, hit the road for Seattle in the cargo space of a blue Jeep Cherokee, accompanied by armed officers in a pair of Washington State Patrol vehicles, lest ski-masked anthro-terrorists stage a daring daylight raid on the convoy as it crested Snoqualmie Pass on its way from the Tri-Cities.

No such excitement ensued. Indeed, K-Man's reception, when he rolled into the parking lot next to the Burke Museum loading dock a little after noon, was indicative of his fading status as a celebrity. Only members of local media were present, and not a great many of them. The Departments of Interior and of Justice sent representatives, but they remained resolutely uncommunicative, suggesting that the press delay asking its questions until the actual scientific examination of the remains began. McManamon himself was present to field questions, but he looked so tweedy and harmless and explained the government's position so quietly and meticulously that TV reporters soon turned from him in search of more colorful material.

The Asatru Folk Assembly provided what photo ops there were, posing for the cameras in their homespun hobbit outfits and chanting at the sky in what was presumably Old Norse. Native Americans were present to conduct a ceremony of welcome and purification, but they failed utterly to consider the media's requirements and conducted their ritual in the privacy of a tent set up for the purpose. None of the plaintiffs appeared for interviews. All in all, a thoroughly pedestrian occasion.

Which, of course, was exactly what the federal authorities had intended it to be. By proposing a detailed, deliberate, phased program for examination of the remains, McManamon and his Interior Department colleagues were depriving the K-Man bonfire of its fuel. Conspiracy theorists can construct an argument from anything, but to engage the imagination of the general public, they must first engage the attention of the media.

True, charges of a government cover-up are always intriguing to reporters, but it was hard to make a convincing case for a cover-up when instead of stonewalling, the government now was going out of its way to call a press conference; handing out a ten-page, single-spaced description of what was going to be done, who was going to do it, where, and why, and inviting questions. The Kennewick bones might still be controversial, but they were gradually ceasing to be news—just as, from a real newshound's point of view, they were finally beginning to get some flesh on them.

*A*part from the staff of the *Tri-City Herald*, few of the journalists covering the Kennewick Man story were reporters, professionally habituated to extracting nuggets of facts from recalcitrant sources. Science writing, feature writing, and editorial writing are all honorable branches of the trade, but they do not require or even reward the kind of devotion to the accumulation of possibly irrelevant detail that distinguishes their colleagues on the sports, city hall, and police beats. The feared and lionized practitioners of the electronic media are in fact even worse off, because most of the words and pictures they live by derive from events staged specifically for their convenience.

A crime reporter assigned to the Kennewick case would soon have been licking her lips and reaching for the phone. But no crime reporter was assigned, and as luck would have it, though the evidence that would have set one sniffing was abundantly available in the court record, neither party to the case was inclined to call the press's attention to it.

The primary evidence in question emerges from comparison of two documents. The first is the description of the Kennewick remains belatedly submitted to the Corps of Engineers on September 11, 1996. This description is undated, but was presumably completed while Dr. Chatters still had the bones laid out in his basement—completed, in other words, before Coroner Johnson took possession of them at the behest of the Corps on the morning of Friday, August 30, 1996. Although this document has been referred to as an inventory, it does not contain a list of all the individual bone fragments discovered during a month of desultory examination of the find site. It is devoted in the main to a list of measurements—bone lengths, diameters, conformations—supporting Chatters's assertion that Kennewick Man did not resemble living Native Americans in physique. But it is detailed enough to make it clear, if only by implication, which bones were complete enough to draw conclusions from.

The second document is almost the exact complement of Chatters's: a three-page, handwritten bone-by-bone fragment inventory of the remains as they were found to be on September 11, 1996, performed in the presence of numerous witnesses in the Battelle Institute storage locker where the remains had been secured six days before, on September 5. Although the Battelle inventory was purely visual and involved no measurement of individual bone fragments, it is detailed enough to identify all the major components of the skeleton.

In the most casual comparison of the two documents, it is glaringly obvious that bones described in detail in Chatters's survey are not mentioned at all in the later inventory. And, far from being small or ambiguous fragments about which professionals might reasonably disagree, the bones in question not only are among the larger and more easily recognized (even by a lay eye) in the human skeleton, but also are among the two or three most significant scientifically for students of the ethnic affinities of ancient humankind.

The three or four bone fragments in question constitute portions of both of K-Man's femora. The femur is the long bone of the upper leg that, after the skull, is the heftiest bone in the body and, with its knobbly top end resembling the bone in a Christmas ham, one of the most easily recognizable. Chatters's list does not indicate how many segments the femora were found in, but his measurements ("maximum length right femur 470 mm, left 470 mm, maximum diameter head 49.3 mm right, 48.8 mm left ...") make it clear that both bones were complete enough to describe as though they were whole. The inventory of September 11, 1996, just as clearly shows that unless they had been reduced to unidentifiable shards of bone, the upper knobby end of one femur (the right) and the lower section of the other were not present.

Furthermore, the large bone fragments in question are said to be clearly visible in photos and videotapes of the remains taken back in August 1996 while they were still in Chatters's custody. They also are missing from the exhaustively detailed inventory of the remains made with Chatters's assistance by the Smithsonian's Owsley prior to their transportation from Richland to the Burke Museum.

If the bones could have walked at any time, even Sherlock Holmes might despair of ever tracking down the guilty party. Fortunately for investigators, however, the window of opportunity was not wide open. There they are in pictures taken in Chatters's basement at the end of August; there they aren't in the inventory made in early September: a period of about two weeks at most. But in fact the window was even narrower than at first appears.

According to Coroner Johnson, he and he alone was present when his deputy coroner placed the Kennewick remains, already enclosed in Ziploc bags, in a plywood box with a matching wooden lid, and then secured the lid in place with duct tape. Leaving Chatters's home, Johnson says he drove directly to the Benton County Courthouse across the river in Kennewick, where he placed the box in the "evidence locker" in the parking lot behind the courthouse for safekeeping.

The evidence locker—actually a roomy two-door garage building—contains numerous storage compartments, each labeled with the name of an individual or department. The one assigned to Johnson is about the size of a military footlocker and is located over six feet above the concrete floor; it is impossible to add to or remove its contents without a stepladder. Like most of the compartments, Johnson's is fitted with a painted plywood door bearing a hasp so that it can be padlocked shut, but both door and hasp are of such flimsy, hardware store quality that few officers bother with them. Rather, they depend for security on the evidence officer in charge of the locker, who logs in all visitors to the building and remains with them for the duration of their visit. So easily visible and awkwardly placed is the locker that it is vanishingly unlikely that anyone, let alone any unauthorized person, could gain access without attracting their chaperone's attention.

There the bones remained over the long Labor Day weekend. On Tuesday, September 3, Chatters, Johnson, county prosecutor Andy Miller, and representatives of the Corps and the tribes met at the Courthouse and agreed that until the controversy over their ownership was resolved, the remains could be stored at Battelle (now the Pacific Northwest National Laboratories). The next day, the boxed remains, now additionally sealed with red "evidence tape" to prevent tampering, were convoyed from the sheriff's evidence locker to building Sigma V on the PNNL campus in North Richland, not far from Chatters's home. There Paul Nickens, one of Chatters's successors as director of the Hanford Cultural Resources program, admitted the cortège to the (electronically secured) building, conducted them to the (locked) room known as "the laboratory," opened a storage cabinet, watched as the (still sealed) remains were placed within, and locked the cabinet again.

According to Nickens and his assistant Mona Wright—the only people with access to keys to the locker and the cabinet—no one visited the room between the time the bones were put there and the "ceremony" six days later when the remains were, long unbeknownst to any but the few individuals present, fully inventoried by an independent expert, though under pressure and hastily, for the first time.

*T*o this day Jeff Van Pelt is not certain why he called his assistant that Tuesday morning and told her to join him at Battelle and to bring her notebook. Suspicion of anything with the name Jim Chatters attached certainly played its part: "At that point it seemed like it wouldn't be long before we'd be burying what was in that box, and I wanted to make sure before we did that we were not burying nothing but a box of rocks." That, at least, was the reason he gave the reluctant Umatilla tribal elders for opening the container, rather than simply praying over it and asking for patience and good will from a troubled spirit for descendants trying to do their best for him.

But you don't need the services of a professional osteologist to distinguish rock from bone or even, for one of Van Pelt's experience, human bone from animal bone. Indeed, animal bone was Julia Longenecker's strong suit; since her marriage, she'd often provided mail-order identification service for archeologist colleagues eager to distinguish chipmunk from squirrel or bovine from ovine bones discovered at their digs. But since returning part-time to active field work as archeological consultant to the Umatilla at the Native American cemetery site discovered in Kennewick by developers of a new golf course, Longenecker had proved that her time spent with one of the greats of the field of human osteology, George Gill of the University of Wyoming, had not been wasted. And besides, no one else was available on such short notice.

While the elders performed a private ceremony in the Kennewick Man storeroom, Van Pelt explained to Longenecker what he wanted her to do. To minimize offense to his fellow Native Americans, Longenecker wouldn't be allowed to touch even the bags containing the fragmentary remains: Van Pelt would remove each bag from the storage box and hold it up for Longenecker's inspection. Only if it was impossible to tell in this fashion what a particular bag contained would he open it and remove the fragment for closer examination. In any event, it did not often prove necessary to do so. Only three of the sixty-eight citations in Longenecker's three pages of notes are marked as unidentified. The remainder, down to quite small portions of ribs and vertebrae, are clearly described: skull, jaw, vertebrae, pelvis, limbs, digits.

After the remains were repacked and reconsecrated, they were once more put under double lock while the observers, including archeologists from the tribes and the Corps, left the building. Van Pelt took charge of Longenecker's notes and returned to the Umatilla Reservation with them, where they were added to the growing file on Techamnish Oytpamanat (The Ancient One). And Longenecker went back to work, with one more adventure in the lively art of cultural resource management under her belt.

Why did it take the better part of two years for someone to compare the inventory made that day by Julia Longenecker under Jeff Van Pelt's supervision with the list and photographs made in Jim Chatters's basement? Probably because it never occurred to anyone that there could be a reason to do so. But once made, all concerned were faced with a real-life version of one of the classic plot lines of the traditional English murder-mystery: "the locked room."

From the time of Conan Doyle's Sherlock Holmes to the present, there are really only two variations on this plot line: Either somebody *did* get into the box, despite locked doors, registered keys, tamperproof tape, and all; or the "missing" bones were never in the box in the first place. In mystery stories, either alternative is equally likely; in real life, where probability, opportunity, and motive count for more than elegant plotting, anyone might be forgiven for thinking that the odds lean pretty strongly toward the second. Chatters's critics would like to see him asked for a full account of his actions during the crucial days before the remains were turned over, this time under oath.

They may yet have an opportunity: At least two of the people who had access to the remains have been minutely questioned by the FBI since spring 1999. But the glacial silence maintained for the better part of a year on the subject by the United States Attorney's office in Spokane, Washington, suggests that the "Case of the Missing Femur Fragments" is not of paramount interest to Federal authorities.

Proponents of justice in the abstract may fume about this passive policy, but it's easy to see why it's been adopted. Since taking charge of the handling of the remains, the Department of the Interior and its point-man Frank McManamon have concentrated every effort on draining the Kennewick Man issue of drama, not pumping up the volume. And in large measure they have achieved what they sought. Despite ferocious innuendoes of scientific incompetence and political bias from the *Bonnichsen* plaintiffs, it was Interior that named the team to evaluate the bones at the Burke Museum in Seattle in February 1999; Interior that in October published a relentlessly circumstantial report on the February studies by the independent panel put together by McManamon: Joseph Powell and Jerome Rose (osteology), Gary Huckleberry and Julie Stein (stratigraphy and sedimentology), ands John Fagan (lithic analysis).[1]

The 64-page report contains only minor differences from Chatters's initial rundown. Basing their opinion on the undamaged condition and near-completeness of the skeleton, Powell and Rose suggest that Kennewick Man was deliberately buried. To Fagan, examination of what could be seen of the stone fragment in the pelvis in X-rays and CAT-scans suggested a somewhat later date—between 7000 and 5000 years ago—but not so conclusively as to overrule the radiocarbon date.

But despite the neutral tone of the report, there was a great deal in it for an ironist to savor. To no one's surprise, the scientists found that the evidence available on surface examination was insufficient to answer the primary question posed them: To which modern-day ethnic group, if any, should the remains of Kennewick Man be repatriated? Further "destructive testing" like radiocarbon and DNA analysis, would be necessary to fulfill the government's duty both to science and to Native Americans.

Negative reaction to the government report from the plaintiffs' party was immediate, but not, this time round, amplified in the media. Even before the Burke team's results were published, Dr. Chatters had learned enough about their conclusions to file a 1700-word affidavit with the court inveighing against its every assumption, procedure, and conclusion,[2] but unlike earlier polemical efforts, this one did not earn Chatters further notoriety with the general public.

Measured strictly in terms of column inches and broadcast minutes, Chatters had already done pretty well out of his brush with celebrity, climaxing in a November 1998 profile in *People*, complete with portrait with wife and dog.[3] Careerwise, however, the proprietor of Applied Paleoscience was learning that celebrity has its downside. Although eminently qualified for the job by training and experience, he failed to secure a tenured teaching position at Central Washington University. His contact with leading members of the profession, despite his continuing involvement in their lawsuit, diminished rather than deepened with time. He was not invited to take part in the public symposium held at the Burke Museum in October 1999 to give a public airing to the political, ethical, and scientific issues raised by the controversy. Nor, though he attended, was he among the speakers at a two-day seminar held a week later in Santa Fe, New Mexico, entitled "Clovis and Beyond."

Although set up on paper like a standard scientific conference on a specialized subject, there was more—and less—to "Clovis and Beyond" than appeared on the surface. Organized by Bonnichsen's Center for the Study of the First Americans, the conference was both an effort to claim reclaim credibility within the profession (lost in the unseemly catfights in court between scientists and government) and to raise funds for the plaintiffs' continued litigation.

As everyone who becomes involved in a pro bono lawsuit quickly learns, there are expenses that no amount of lawyerly good will can avoid: for services of writs, filing of documents, faxing and phoning, and reams upon unimaginable reams of copying. By December 1998, a group calling itself Friends of America's Past had turned up on the World Wide Web, dedicated to "promoting and advancing the rights of scientists and the public to learn about America's past," and, more concretely, "to raise funds to support the Kennewick Man case." It was hardly surprising to discover that the Friends had the same address as the *Bonnichesen* plaintiffs' law firm, and that their registered agent was *Bonnichesen* attorney Alan Schneider.

As a fundraiser, "Clovis and Beyond" scored: over a thousand registered for the conference at a minimum $100 a head. On the credibility side, however, little ground was gained. The two-day run of the conference just before Halloween 1999 offered a series of straightforward scientific presentations—mostly surveys of research already published—bookended by "panels" of sound-byte-sized statements on the future of science and public policy in early-American research and a whirlwind tour (10 minutes per speaker) of notions for future exploration of the subject.

Despite the eminence of many of the speakers on both panels, among them Monte Verde discoverer Tom Dillehay and Folsom re-excavator David Meltzer, the overall impression left by the conference was one of covert strife masked by barely maintained professional politeness. Characteristic of the edgy tone of much of the discourse was the conference's very first statement, at Friday's "The Future of Public Policy: How Do We Go From Here?"

After a polite nod to his host, Rob Bonnichsen, the current president of the Society for American Archeology, Keith Kintigh of Arizona State University, proceeded to throw down the gauntlet on the question that, though unspoken, had drawn many of the curious among the conference audience. "It is the position of the Society for American Archeology," said Kintigh, "that under NAGPRA, 'First Americans' *are* Native Americans, regardless where they came from, when they came, how many migrations there were, or whether some groups among them died out."

To the extent that science was its objective, "Clovis and Beyond" gave little comfort to the partisans of "caucasoid" priority in North America, or the kind of bizarre ethnic theorizing about the origins of the first Americans so tersely ruled out of bounds by Kintigh. As he had in Seattle the week before, the University of New Mexico's Powell spoke eloquently about the dubiousness, even absurdity, of reading contemporary racialist messages into the results of osteometric work on ancient remains.

So far as the mass media were concerned, he might as well as not have spoken at all. Chatters's Captain Picard bust of Kennewick Man may have lost some of its intial credibility, but other such renderings, marginally more scientific, have joined the paleo-American portrait gallery. An early November story in the *New York Times*[4] featured not only K-Man (in a new, improved version marked with felt-tipped-in worry lines), but a bust of the more or less contemporary Spirit Cave Mummy from Nevada, as well as a wildly speculative reconstruction of an individual supposedly dating to 11,500 years ago asserted by its discoverer as clearly more closely related to ancient Australasians or Southeast Asians than any geographically more plausible group of the time.

Not on the conference program proper was a reprise by the Smithsonian's Dennis Stanford of his thesis that resemblances between early stone tools in the Eastern United States and Western Europe demonstrate that North America might well have been settled via a trans-Atlantic, not a circum-Polar route. As after-dinner entertainment for attendees at the closing banquet of "Clovis and Beyond," the speech went down easily. This, too, though almost universally regarded by specialists as nonsense, was reported without comment by *The New York Times*.

At press time for this book, the Department of the Interior was awaiting results of four further radiocarbon samples taken from bones in the Kennewick Man skeleton. All indications are that the dates will fall in the very near neighborhood of that already recorded: roughly 9000 solar years ago. Given the reported condition of the bones, it is not terribly likely that enough intact organic matter remains in them to perform meaningful DNA analysis should the government and the courts demand it.

Even if analysis is performed and produces comprehensible results, it would not serve to answer the questions that many members of the general public and press are asking. By now, it no longer matters that many of these questions are, scientifically speaking, all but meaningless. Kennewick Man and his less-publicized ilk have long ceased to be elements of evidence for rational argument, but rather have become characters in a drama, a tabloid fantasy of the past, in which probability counts little when weighed against colorful speculation, and truth is routinely ignored in the search for a good story.

Epilog: The View from Jump-Off Joe

*I*t's a good twelve miles as the crow flies from the Columbia racecourse to the top of Jump-Off Joe. You'd never guess, this August Sunday afternoon, that 30,000 or 40,000 people are patiently sweltering down there along the river, watching Dave Villwock and Chip Hanauer duke it out for the Columbia Cup. But the sound of the hydros penetrates even here through the shimmering air, faint but threatening, like the sound of angry wasps trapped in a jam jar.

Nobody much comes to Jump-Off Joe except the technicians in charge of upkeep for its forest of microwave antennas and broadcast towers. The thin litter of broken brown bottle-glass among the tufts of thin grass between the basalt outcrops testifies to the most frequent visitors: kids in borrowed pickups with a case of beer in the ice chest and a need to get away from the dust of the flatlands for a while. Mention Jump-Off Joe in the Tri-Cities area, and most adults will tell you they've never been up, wouldn't even know how to get here. Always meant to, but somehow never got around to it.

Devotees of scenery with a capital S might be disappointed by the vista. The predominant foreground color is pale beige with intermittent streaks of near-white where the rivers have cut deep into the dusty soil, and seams of dark rock where the soil has been scoured away completely. On a clear day you might glimpse the snowpeaks of the North Cascades to the northwest and the Blue Mountains looming low on the southwest horizon. The trouble is, really clear days are rare. In plowing season, plumes of dust rise a thousand feet into the air behind every tractor, after-harvest stubble burns send up a dark pall, and west winds lift the fine dust that carpets the plateau lavas and shift it toward the fertile dust mountains of the Palouse country, as they have done ever since the glaciers rimmed this country round. Even when the air is quiet, an almost invisible haze hangs above the course of the river, thickening to mist in the coldest days of winter.

But all the most notable aspects of the landscape are more accessible to the mind's eye than to the Instamatic lens anyway. Along a north–south line a little to the east of here, for example, if you had Superman's X-ray vision, you could look ten or fifteen miles deep into the earth and study what used to be the western margin of North America, before a quarter-billion years' worth of random eruptive detritus and oceanic muck began to accumulate layer upon layer against the coastline. (Bits of the Blues may date back to the dawn of life on earth, but their rocks didn't raft up against the mainland until much later, well into the Age of Dinosaurs.)

Most of what you see from here, including the stones beneath your feet, came along pretty late in the process, starting in the mid-Miocene some twenty million years ago, when the dinosaurs were already long gone and our mammalian ancestors had inherited their territory. The same forces crushing, twisting, stretching the coastline helped to open cracks in the continent off to the south and east, releasing hot, deep, liquid rock that flowed across the landscape like water. These floods of lava continued intermittently for about 10 million years, until the entire 250-mile-wide basin between the mountains of east and west was filled, a mile and more deep.

Intermittently, that is, geologically speaking: After about the first twenty catastrophic outbreaks, the lava flooding settled down to every half-million years or so—plenty of time for things to cool down, for the rivers to begin to find their way back to the sea across the plain, for dust to blow into the crannies, and for life gradually to return, only to be wiped out to the last blade of grass by the next superheated inundation.

The last big outbreak came a good six million years ago and never got quite this far, but geologically, conditions remained anything but tranquil. When the liquid lavas ceased to spew, the volcanoes rising to the west began pumping out clouds of ash, some so fine and vast that the winds carried them as far as Montana and Wyoming. After a few million years of desultory fire came the ice. The Columbia plateau was spared by the glaciers as they advanced and retreated during their two-million-year campaign, but their neighborhood, combined with the rising mountains to the west blocking access to the moist Pacific breezes, turned the area's humid, subtropical climate to chilly semi-desert scoured by dry cold winds off the ice.

It was the ice, indirectly, that produced most of the soil that came to blanket the lava plains, and even more indirectly, it was the ice that took the soil away again. Glaciers do a pretty good job of grinding bedrock to powder as they move, and the rock powder steadily gets sluiced out from beneath, washes up on banks of meltwater streams, and blows away with the wind. In glacial days the wind came from the west, just as it does today, and carried the powdery rock-dust hundreds of miles from its source, to collect in soft mounds hundreds of feet deep against the western flanks of the high plains.

A lot of that dust is still in place today. You can see it whenever farmers plow in the Palouse country northeast of here where winter wheat is grown. The soil looks tasty enough to eat with a spoon: a rich, dark, chocolate-pudding brown. Not a trace of it remains here. Starting around 15,000 years ago, when the climate was once more teetering toward warmer conditions, a huge meltwater lake formed in a broad valley in the Rocky Mountain foothills. Unfortunately for the countryside to the west, the only thing keeping that lake in place was a dam of ice, and when the dam was breached, the entire contents of "Lake Missoula" (try to imagine a cube of water over fifteen miles on a side) headed for the Pacific all at once.

Some geologists describe the Lake Missoula floods as the most catastrophic in the planet's history. And there were more than forty of them over the two or three thousand years it took for the ice to pull back from western Montana entirely. By the time they ended, a swath of eastern Washington a hundred miles wide had been scoured clean of topsoil—and not just scoured, but carved into gullies miles wide and hundreds of feet deep, cut in solid bedrock. The only soil you can see over most of the vista from Jump-Off Joe was left behind by those floods, settling out in the thousand-foot-deep lake that drowned the landscape until the waters could leak downstream through the narrow hairpin gap where the Columbia cuts through the very ridge you're standing on.

Then, after some 15 million years of abuse by fire and ice, things calmed down a little. By the time Kennewick Man and his kind arrived 10,000 years ago, the worst they had to cope with geologically was the occasional rain of ash from a Cascade volcano. (Some of those events were bad enough to turn the area into a temporary dead zone. K-Man's bones lay beneath a layer of ash from the 4900 B.C. explosion of Oregon's Mt. Mazama, which left traces as far east as Wyoming).

This has never been a country friendly to humankind. Few edible plants are native here, and few game animals—no big game at all nearer than the Okanogan highlands or the Blue Mountains. It was the incredible, recurring miracle of the salmon that brought people to this arid plateau, whether from the high country to the east, as the Clovis-first crowd believes, or up the Columbia, as in the coastal-settlement hypothesis. But even the salmon weren't enough to support more than a few widely scattered villages year-round. Most of the people who left their artifacts, their arrowheads, and sometimes their bones along the banks of the broad river below were summertime visitors. It wasn't until the white settlers came and began to harness the river itself that settlements big enough to see with the naked eye took root.

Of course, that is the story according the white settlers and the shamans of their tribe. Some Native Americans put forward a simpler scenario. Wherever the interlopers came from, they say, *we* were born from the earth, are children of the earth. They don't have to draw up a family tree to establish their fraternity with Kennewick Man, no matter how many thousands of years have passed since his death. By his very presence in this land, he is their brother, whereas those who re-fuse him return to the earth are grave robbers, whatever rationale they for their crime.

Such Old Believers are very much in the minority, probably even among Native Americans. The thing that upsets many, even those whose prophets are Galileo and Darwin and Nietzsche, is the utter failure on the part of the Old Ones of science to acknowledge that their world view, too, is ultimately based on faith.

"Ye shall know the truth," says Jesus in the Gospel of St. John, "and the truth shall make you free." If any dictum lies at the root of the scientific enterprise, it is this. But the tense of the verb in question is future. *Knowledge* of truth lies always in the future; all we can do the present is press onward toward it. A claim to possess it already is fatal arrogance, antithetical to the search. One of the wisest living of the white tribe, writing of the boundary between what can be known and what can be said, put it this way: "Everything is only a metaphor. There is only poetry."[1]

Out there, beyond the scatter of human occupation along the river, the White Cliffs of the Columbia (silts laid down by the Lake Missoula floods) gleam pale in the westering sun. By now, the freeway must be choked with cranky, sunburned people driving home to Sunday dinner. That's invisible, inaudible from here. The only sound, now that the thunder-boats have fallen silent, is the stridulate chirr of crickets in the grass. The prospect from up here is grand and inspiring, but lonely. It's good from time to time to get away, in fact or metaphor, to take the long view, but soon it's time to return to the lowlands. Time now, as the poet says, to go: "... human kind cannot bear very much reality."[2]

NOTES

CHAPTER 2

1. Quoted in the *Seattle Times*, June 10, 1998

2. The only fairly complete account of the Marmes dig is that of Ruth and Louis Kirk, *The Oldest Man in America: An Adventure in Archaeology* (New York: Harcourt, 1970), which has long been out of print, but is worth tracking down. An official report on the dig has never been published, though the voluminous materials collected remain accessible to scholars at Washington State University.

3. "Evolutionary Human Paleoecology: Climatic Change and Human Adaptation in the Pahsimeroi Valley, Idaho, 2500 PB to the Present," 1982: unpublished thesis).

4. All excerpts from Chatters's "field notes" were transcribed by the author from copies provided by the U.S. Army Corps of Engineers, Walla Walla Region. Abbreviations (inch and foot symbols, ampersands, and the like) have been silently expanded. Expansions of abbreviated words are indicated by [] brackets.

5. Quoted from Tracy's declaration of April 21, 1997, submitted as a memorandum in the case *Bonnichesen et al.* v. *United States of America et al.* in the U.S. District Court for the District of Oregon, case no. CV 96 1481 JE.

6. Emphasis added.

CHAPTER 3

1. The text of this e-mail, which played a major role in defining the ground upon which the later legal battle over the remains would be fought, follows in its entirety.

> **From:** Richard L Jantz
> **Date:** Sat, 28 Sep 1996 09:26:09 -0400 (EDT)
> **Subject:** Paleoindian remains (fwd)
>
> Recently a 9000 year old skeleton was discovered near Richland, Washington on federal land managed by the U.S. Army Corps of Engineers. The Corps has decided to surrender custody (repatriate) this skeleton to local Native American tribes without formal study. The local district engineer (LTC Curtis) has denied repeated requests to examine and measure the remains.
> This is an extremely important skeleton because of its antiquity and the impact its loss would have on the entire repatriation process, i.e. immediate return of very ancient bones without thorough documentation.

The Corps made its decision without the benefit of any formal report. The remains were briefly examined by James Chatters, Grover Krantz and Catherine MacMillan. All three observed features that were atypical of modern Native Americans and typical of features more frequently seen in modern Caucasians.

Recent work by Gentry Steele has observed that early American crania exhibit greater similarity to Europeans and Southern Asians than they do to modern Native Americans. It cannot automatically be assumed that early American skeletons are ancestral to modern Native Americans. If a pattern of returning these remains without study develops, the loss to science will be incalculable and we will never have the data required to understand the earliest populations in America.

We have less then 30 days before the remains are turned over. If you wish to express your displeasure with the Corps' action, please write or fax.... Thanks for your help in this matter....

2. Quoted in the *Tri-City Herald*, October 3, 1996

CHAPTER 4

1. Including the textbook *The Stages of Human Evolution*, 5th ed. (1994) and 1979's *Atlas of Human Evolution*.

2. Transcript of motion hearing, October 23, 1996, CV 96-1481 JE, docket #29, p. 16.

3. The following material draws heavily on the researches of Stephen Jay Gould, whose monograph *The Mismeasure of Man* (New York: Norton, 1981; rev. ed. 1996) is the most powerful treatment to date of how the conclusions of a science can be conditioned and distorted by the unexamined mindset of its practitioners.

4. Gould, *Mismeasure*, p. 46.

CHAPTER 5

1. The following is based on David Meltzer's own words recorded in July 1998 at the Folsom site, supplemented by Meltzer's fine book about the development of the field of North American prehistory, *Search for the First Americans* (Washington, D.C.: Smithsonian Books, 1993).

2. McJunkin, a former slave, deserves a book of his own, and has one in *Black Cowboy: The Life and Legend of George McJunkin* by Franklin Folsom (Niwot, CO: Roberts Rinehart, 1992). A shorter survey of McJunkin's remarkable career can be found in "Fossils and the Folsom Cowboy" by Douglas Preston, in the magazine *Natural History*, February 1997, p. 16.

3. In his *Notes on the State of Virginia*, written in 1781 and published 1787. The relevant passage can be found on pp. 218–228 of the Library of America edition of Jefferson's writings.

4. There are any number of books that recount the prehistory of North American anthropology in greater or lesser detail. For this, as for most background information, I have most often referred to Brian Fagan's survey volume *Ancient North America: The Archeology of a Continent* (London: Thames and Hudson, 1995).

CHAPTER 6

1. Any number of writers interested in pseudoscientific arcana have dipped into this rich literature and extracted tasty morsels. One of the most exhaustive treatments of the phenomenon is Stephen Williams's *Fantastic Archeology: The Wild Side of North American Prehistory* (Philadelphia: University of Pennsylvania Press, 1991), though at more than 400 pages it may be too much of a good thing for any but specialized readers.

2. Hannah Marie Wormington, *Ancient Man in North America*, 5th ed. (Denver, CO: Denver Museum of Natural History 1964) p. 3.

3. Breternitz *et al.*, "An Early Burial from Gordon Creek, Colorado," *American Antiquity* 36(2): 171–181 (1971).

4. A lively, not to say lurid, account of Turner's highly controversial theories about the end of the Chaco culture by thriller writer Douglas Preston is to be found in "Cannibals of the Canyon," *The New Yorker*, November 30, 1998.

5. C.G. Turner, "Three-rooted Mandibular First Permanent Molars and the Question of American Indian Origins," *American Journal of Physical Anthropology* 34:229–241 (1971). Turner's subsequent publications on the subject are legion. One of the more accessible is "Relating Eurasian and Native American Populations Through Dental Morphology" in Bonnichsen and Steele's *Method and Theory for Investigating the Peopling of the Americas* (CFSA, Oregon State University, 1994), an invaluable overview of the subject.

6. In a paper presenteed at the 5th International Conference of Anthropological and Ethnological Sciences in 1956 and published in 1960 by the University of Pennsylvania Press in a collection entitled *Men and Cultures*, edited by Anthony F.C. Wallace.

7. J.H. Greenberg, *Language in the Americas* (Stanford, CA: Stanford University Press, 1987).

8. "The History and Classification of American Indian Languages: What Are the Implications for the Peopling of the Americas?" in Bonnichsen and Steele 1994, *op. cit.*, pp. 189–207. The same collection includes a vigorous defense of Greenberg's tripartite analysis by Merrit Ruhlen of Greenberg's old department at Stanford (pp. 177–188).

CHAPTER 7

1. The original paper by Krings et al. was published in 1997 in *Cell* 90: pp. 1–20. A clear exposition of the results and their background appears in Patricia Kahn and Ann Gibbons, "DNA from an Extinct Human," *Science*, July 11, 1997, pp. 176–178.

2. M.F. Hammer, "A Recent Common Ancestry for Human Y Chromosomes," *Nature* 378:376–378. By far the best substantial overview of recent Y chromosome literature is "The Role of the Y Chromosome in Human Evolutionary Studies," by Hammer and his anthropologist colleague at UA Stephen Zegura, published in 1996 in *Evolutionary Anthropology* 5/4:116–134.

3. In Cavalli-Sforza's case, "the book" is *The History and Geography of Human Genes* (Princeton, NJ: Princeton University Press, 1994). A paperback edition ("abridged" by omitting hundreds of pages of gene-frequency tables, but still weighing in at over 400 double-column, small-print pages) was published in 1996. Despite some tough math and impenetrable paragraphs here and there, it's a must for the library of anyone seriously interested in the theory and practice of biological prehistory. Far easier to read as a general introduction to the subject is Cavalli-Sforza's quasi-autobiographical collaboration with his documentary filmmaker son Francesco, *Chi Siamo*. This very interesting account, the title of which is *Who We Are*, has been published in English translation as *The Great Human Diasporas* (Reading, MA: Addison-Wesley, 1995).

4. Beginning with one in the June 1992 issue of *Human Biology*, "Paleobiology of the First Americans," pp. 138–146.

5. The best single widely available source for Powell and Steele's studies is D.G. Steele and J.F. Powell, "The Paleobiology of the First Americans," *Evolutionary Anthropology* 2/4:138–146 (1993).

CHAPTER 8

1. For the bad news—and the good news—about the role of extramarital sex in keeping the gene pool stirred up, read "The Science of Adultery," Chapter 4 of Jared Diamond's marvelous survey of human evolution, *The Third Chimpanzee* (New York: HarperCollins, 1992).

CHAPTER 9

1. *Tri-City Herald*, July 27, 1997.

2. Jonathan Mazzochi; "Race and Relics," *The Dignity Report* 5/1:4–7.

3. "The Buhl Burial: A Paleoindian Woman from Southern Idaho," *American Antiquity* 63/3:437–456 (1998).

4. Samantha Silva, "A Famous Skeleton Returns to the Earth," *High Country News*, March 8, 1993.

5. Douglas Preston; "The Lost Man," *The New Yorker*, June 16, 1997.

CHAPTER 10

1. *Tri-City Herald*, July 31, 1997.

2. *Tri-City Herald*, June 16, 1998: "British TV wants tiny favor: a few hydros on Columbia."

3. Francis P. McManamon, "Approach to Documentation, Analysis, Interpretation, and Disposition of Human Remains Inadvertently Discovered at Columbia Park, Kennewick, WA," National Park Service, Department of the Interior, November 1998.

4. Draft report dated June 29, 1998.

CHAPTER 11

1. Affidavit of James Chatters submitted in *Robson Bonnichsen et al.* v. *United States*, in support of the motion to quash subpoena and for a protective order. Docket no. 72, dated March 31, 1997.

2. David J. Meltzer: "A Review of *Monte Verde and the Pleistocene Peopling of the Americas*, vol. 2," *Science*, May 2, 1997, 754–755.

CHAPTER 12

1. See R.G. Matson and Gary Coupland, *The Prehistory of the Northwest Coast* (New York: Academic Press, 1995) for a full-dress reconsideration and ratification of the theory.

2. K.R. Fladmark: "Routes: Alternative Migration Corridors for Early Man in North America," *American Antiquity* 44/1:55–69 (1979).

3. *Archeological Investigations at Early Village Sites in the Middle Santa Cruz Valley: Analyses and Synthesis*, ed. Jonathan B. Mabry, Anthropological Papers No. 19, Desert Archeology, Inc., 2 volumes, 1996.

4. "A Modern-Day Inquisition by the Pyramid Lake Paiutes," *Nevada State Museum Newsletter*, 26/2 (March/April 1998): 4–5.

CHAPTER 13

1. The full report is available on the Cultural Resources website of the National Park Service, http://www.cr.nps.gov/aad/kennewick/.

2. Affidavit dated 14th day of September, 1999. Available on the web at http://www.friendsofpast.org/affidavit-chatters-991001.html.

3. "Head Case," *People*, November 30, 1998, pp. 181-182.

4. "New Answers to an Old Question: Who Got Here First?" by John Noble Wilford, *New York Times*, November 9, 1999.

EPILOG

1. Concluding words of *Love's Body* by Norman O. Brown (New York: Random House, 1966).

2. T.S. Eliot, *Four Quartets*, "Burnt Norton," I:43–44. In *Collected Poems* (New York: Harcourt), p. 118.

INDEX

W

Washington State University, Pullman, Washington (WSU), 14–17
Whitewater Draw, Arizona (early-remains site), 83
Wormington, Hannah Marie (curator, Denver Museum of Natural History)
    author of *Ancient Man in North America*, 82
    early human specimens listed in 5th edition (1964), 82

Y

Y-chromosome DNA, use in tracing evolution of human diversity, 98–99

GAYLORD R